# THE PROGRAM
# EVALUATION PRISM

# THE PROGRAM EVALUATION PRISM

## Using Statistical Methods to Discover Patterns

Martin Lee Abbott

Department of Sociology

A JOHN WILEY & SONS, INC., PUBLICATION

For general information on our other products and services or for technical support, please contact our Customer Care Department within the United States at (800) 762-2974, outside the United States at (317) 572-3993 or fax (317) 572-4002.

Wiley publishes in a variety of print and electronic formats and by print-on-demand. Some material included with standard print versions of this book may not be included in e-books or in print-on-demand. If this book refers to media such as a CD or DVD that is not included in the version you purchased, you may download this material at http://booksupport.wiley.com. For more information about Wiley products, visit www.wiley.com.

*Library of Congress Cataloging-in-Publication Data:*

Abbott, Martin Lee
    The program evaluation prism : using statistical methods to discover patterns / Martin Lee Abbott.
        p. cm.
    Includes bibliographical references and index.
    ISBN 978-0-470-57904-6 (pbk.)
    1. Educational evaluation–United States–Statistical methods.  2. SPSS (Computer file)  I. Title.
    LB2822.75.A23 2010
    379.1'58021–dc22

                                                                                    2009051018

10 9 8 7 6 5 4 3 2 1

# CONTENTS

**Preface**      xi

**Acknowledgments**      xiii

**1   Introduction**      1

Initial Considerations / 4
Book Plan / 5
Real Examples / 6
     Using Statistical Programs / 7
     The Evaluator's Journey / 7

**2   The Elements of Evaluation**      9

The Nature of Evaluation / 9
Evaluation Concerns / 10
Evaluation Standards / 12
Methods used in Evaluation / 12
The Evaluator's Tools / 13
Evaluation Hurdles / 14
     Quantification / 14
         *Resistance to Quantification / 15*
         *The Nature of Quantification / 15*
         *Qualitative Methods / 17*
     Specialization / 18
Statistical Issues / 19
     Certainty versus Probability / 19
     Statistical Significance / 20
     Effect Sizes / 20
     Can We Achieve Certainty? / 21

Dispelling the Mystique of Statistics / 22
Research Literacy / 23
The Discovery Questions / 24
School Characteristics and Student Learning / 24
Worker Participation / 25
The Impact of Technology on the Classroom / 27
Classroom Observation Data / 28
Discovery Learning / 29
Terms and Concepts / 29

**3   Using SPSS®**                                                     **31**

General Features / 32
Management Functions / 34
Reading and Importing Data / 34
Sort / 34
Split File / 35
Transform/Compute (Creating Indexes) / 36
Merge / 37
Analysis Functions / 41
Graphing Functions / 41

**4   Correlation**                                                     **47**

The Nature of Correlation / 47
Prediction / 49
Correlation Is Not Causation / 50
Pearson's *r* / 51
Strength and Direction / 51
A Note on the Nature of the Data / 54
Interpreting Pearson's *r* / 56
Testing the Statistical Significance of a Correlation / 57
The "Practical Significance" of *r*: Effect Sizes / 60
An Evaluation Example of Correlation: The Impact of
   Technology on Teaching and Learning / 63
Influences on Correlation / 71
Restricted Range / 71
Extreme (Outlier) Scores / 73
Other Kinds of Correlation / 73
A Research Example of Spearman's
   rho Correlation / 75

Nonlinear Correlation / 78
"Extending" Correlation to Include Additional Variables / 78
Correlation—Detail for the Curious / 78
  Computing Pearson's *r* / 78
  Assumptions of Correlation / 79
  Nonlinear Correlation / 81
Discovery Learning / 81
  Terms and Concepts / 83
  Real World Lab—Correlation / 84
  Description of the Data / 84
  Evaluation Questions / 86


**5  Regression**                                                              **87**

The Regression Line—Line of "Best Fit" / 88
The Regression Formula / 91
Standard Error of Estimate / 93
Confidence Interval / 94
Residuals / 94
Regression Example with Achievement Data / 98
  The Results of the Analysis / 101
  The Graph of the Results / 105
  Standard Error of the Estimate / 106
  The Confidence Interval / 108
Detail—For the Curious / 110
  Assumptions of Regression / 110
  Fixed and Random Effects Modeling / 111
  Nonlinear Correlation / 111
  Calculating the Standard Error of the Estimate / 116
Discovery Learning / 117
  Terms and Concepts / 118
  Real World Lab—Bivariate Regression / 120


**6  Cleaning the Data—Detecting Outliers**                                    **121**

Univariate Extreme Scores / 122
Multivariate Extreme Scores / 124
Distance Statistics / 128
Influence Statistics / 131
Discovery Learning / 134

Terms and Concepts / 135
Real World Lab—Extreme Scores / 136


**7  Multiple Correlation**                                                137

Introduction / 137
   Control Variables / 139
   Mediator Variables / 140
Using Multiple Correlation to Control Variables:
   Partial and Semipartial Correlation / 142
Partial Correlation / 143
Semipartial (Part) Correlation / 147
Discovery Learning / 149
   Terms and Concepts / 149
   Real World Lab—Partial and Semipartial
      Correlation / 151


**8  Multiple Regression**                                                153

Multiple Regression with Two Predictor Variables / 154
   Uses of Multiple Regression / 154
   Multiple Regression Outcomes / 155
      *Omnibus Findings for the Overall Model / 155*
      *Individual Predictors / 157*
      *Additional SPSS® Results / 158*
Multiple Regression: How to Enter Predictors / 165
Stepwise Regression and Other Methods / 167
Assumptions of Multiple Regression / 171
   Multicollinearity / 171
Cleaning the Database / 174
Multiple Regression with More Than Two Predictor
   Variables: Research Examples / 174
   Predicting the Impact of School Variables on
      Teaching and Learning: The TAGLIT Data / 174
      *Omnibus Findings / 177*
      *Results of Individual Predictors / 177*
The "Larger Model" of School Achievement / 178
Discovery Learning / 180
   Terms and Concepts / 181
   Real World Lab—Multiple Regression / 182

**9  Coding—Using Multiple Regression with Categorical Variables**     **185**

Nature of Dummy Variables / 185
One Categorical Variable with Two Groups / 186
Creating Dummy Variables / 193
Creating Subvariables in SPSS® / 194
One Categorical Variable with More Than Two Groups / 199
  A Hypothetical Example / 200
  An Example from the School Database / 202
Discovery Learning / 203
Detail for the Curious—False Dichotomies / 203
  Terms and Concepts / 204
  Real World Lab—Dummy Coding / 205


**10  Interaction**     **207**

Interactions with Continuous Variables / 209
Interaction with Categorical Variables / 216
Discovery Learning / 221
  Terms and Concepts / 222
  Real World Lab—Interaction / 223


**11  Discovery Learning Through Correlation and Regression**     **225**

Overall Discovery Notes / 225
Findings from the Data / 226
  Student Academic Achievement / 226
  Workplace Participation / 227
  Impact of Technology on Student Learning / 228
Advanced Statistical Techniques / 229
  Hierarchical Linear Modeling / 229
  Structural Equation Modeling and Path Analysis / 230
Other Regression Procedures / 231


**12  Practical Application Analyses**     **233**

Real World Lab—Correlation / 233
Real World Lab—Bivariate Regression / 235
Real World Lab—Extreme Scores / 239

Real World Lab—Partial and Semi-Partial
  Correlation / 245
Real World Lab—Multiple Regression / 247
Real World Lab—Dummy Coding / 249
Real World Lab—Interaction / 252

**Appendix**                                                     **259**

**References**                                                   **305**

**Index**                                                        **307**

# PREFACE

As someone conducting or learning to use evaluation methods, you will be in a position in which you need to use data to understand a research problem. You may have a good intuitive sense of correlational methods but not be formally trained in statistical analysis; your goal is to arrive at sound conclusions as the basis for quality decision-making. If so, this book is for you. Over the years, I have worked with many people in your position and have enjoyed watching their research come alive as they developed the tools contained in this book. My hope is to put the power of evaluation research methods into your hands.

Recently several popular books have been published that emphasize the need for examining data to reveal surprising answers. The authors of the bestselling *Freakonomics* (Levitt and Dubner, 2005) provided a glimpse of using statistical processes in this way, yet the description and richness of the processes for helping people make sense of the problems around them are largely unexplored. More recently Ayres (2007) described how huge databases are mined for information that can be used to shape consumer choices and opinion. Ayres very thoughtfully noted the statistical processes underlying these activities, in particular highlighting regression and randomized trials as tools in "data-based decision-making." Both of these books are excellent introductions to how evaluation processes are used to find patterns that affect daily life. What is missing in most of these attempts is a detailed examination of the methods that will allow evaluators to discover the dynamics hidden in their programs or focus of examination. This is what I hope to provide in this book.

It is important to recognize at the outset that any book discussing statistics, and multiple regression specifically, is written against the backdrop of many important advancements of these complex procedures. This is particularly important for a book that focuses on applying these procedures to real-life evaluation problems. While each advancement contributes to the universal body of statistical understanding, it is important to point to some influences on my own thinking over the years. I acknowledge these sources throughout the book for different aspects of the statistical procedures I discuss. You should seek these out, along with other similar references, for specific questions that go beyond the coverage of this book.

It is my hope to be able to make the statistical concepts and procedures developed by these authorities accessible to the evaluation researcher. By advancing our knowledge, we can improve the procedures we use to provide meaningful results in our evaluation efforts.

I have made a range of resources available to readers that might be useful for developing a deeper understanding of the topics in this book. The Appendix includes two research articles that I wrote along with two of my colleagues at the Washington School Research Center. Both use the methods I discuss in this book. The articles are reprinted here with the permission of the WSRC and can be viewed at the following website address: http://www.spu.edu/orgs/research/currentresearch.html

Please use the ftp address below to access and download the following:

- Large and small data sets for application of the procedures discussed in the book
- SPSS lab exercises
- Access to modules of my online course in graduate research methods
- Access to evaluation research reports that we have published at the Washington School Research Center as examples of using statistics for program assessment and evaluation

ftp://ftp_wiley_com/public/sci_tech_med/evaluation_prism

Additional resources and information will also be available at my personal website:

http://myhome.spu.edu/mabbott

# ACKNOWLEDGMENTS

I would like to thank everyone who reviewed this manuscript. Jeff Joireman's thorough critique was invaluable throughout the process. All of the following people provided thoughtful comments helpful to the direction of the manuscript at various stages: Nathan Brown, Nyaradzo Mvududu, Kari Peterson, and Tom Trzyna. I also want to thank Duane Baker (The BERC Group, Inc.) and Liz Cunningham (T.E.S.T., Inc.) for approval to use their data in this book. Using actual data of this nature will be very helpful to readers in their efforts to understand statistical processes. A special thank you to Dominic Williamson for help with cover design and creating my personal website. Kristin Hovaguimian worked wonders with the Index and Kathleen Abbott's tireless support was indispensable.

I especially want to recognize Jacqueline Palmieri and Steve Quigley at John Wiley & Sons, Inc. for their encouragement and support from the beginning of this project to develop a book with a different approach to the use of statistical methods.

# 1

# INTRODUCTION

My first glimpse of how evaluation methods can illuminate understanding came in studying industrial democracy. At the time it was fashionable to institute worker participation programs to encourage greater ownership of the work process and, it was hoped, greater productivity. Many companies followed suit, and all appeared to be rosy. Management assumed that if they gave the workers a chance to participate in decisions about the conditions that affected their work lives and behavior, they would jump at the opportunity. This would supposedly result in positive outcomes for the organization as well as enrich workers' lives. This expectation proved to be shortsighted, however, and management relations often became difficult. A close analysis of some of these attempts revealed patterns among the attitudes and desires of the workers that were unanticipated. Having the opportunity to participate in decision-making did not always automatically result in the predicted worker attitudes; other attitudes (e.g., whether or not one *desired* to participate) had an effect on the outcomes of the study.

Like light shining on a prism, the power of evaluation can help reveal hidden patterns in data when these data are exposed to the right methods. New understanding can emerge from a careful and disciplined examination of even "mundane" practices that we take for granted. Ultimately, using evaluation methods can help illuminate the world

*The Program Evaluation Prism: Using Statistical Methods to Discover Patterns,*
by Martin Lee Abbott
Copyright © 2010 John Wiley & Sons, Inc.

around us in a different light so that we see different sides of the same realities we face in everyday life. By using careful methods, a company might avoid difficulties and increase the chances for positive outcomes by becoming more aware of the underlying patterns of meaning that influence work.

The power of evaluation methodology is the power of discovery and understanding. Peter Berger (1963) once defined sociological discovery as "culture shock minus geographical displacement," (p. 23) emphasizing how we can be surprised at the seemingly "ordinary" reality we inhabit by looking at it through the lens of the sociological perspective. In the same way, utilizing appropriate evaluation strategies can lead to enlightening discoveries that are "right in front of our faces." Collecting and analyzing data may allow us to make observations and reflections that will cast an entirely different light on the nature of a problem. We may be shocked at what we find because it may be different from our initial expectations, or may reveal new program directions.

My experience after 30 years of teaching and conducting evaluation research is that most practitioners and students do not fully take advantage of evaluation methods and therefore cannot partake of the insights they provide. In the case of evaluators hired by a company to produce research, many may be unfamiliar with research and statistical processes and seek to subcontract parts of the work to "experts" who can help them in their work or who can provide simple analyses that may support their overall observations. In other cases, evaluation work may not be well served by powerful methodological processes because these processes are not well understood or because the appropriate conditions are not present that are required for their use.

In a wider sense, the public misunderstands and avoids evaluation processes for several reasons. "Math phobia" is probably the stock answer given to why most people avoid the statistical tools of evaluation, which make use of numbers and symbols that appear arcane and inscrutable. Statistics also has a bad reputation for being used incorrectly to support whatever the researcher promotes. This popular notion is most readily recognized in the book *How to Lie with Statistics* (1993). To be sure, there continues to be a great deal of abuse when statistical processes are used as divining rods by practitioners who either do not understand the processes or are simply unscrupulous. It is easy to "lie with statistics." But we must also recognize that, in some sense, numbers do not lie; the issue is not the numbers but the interpretation of the meaning of the numbers.

This book explores the nature of evaluation and how evaluation measures can be used to illuminate hidden patterns of meaning.

Like the sculptor's chisel, evaluation tools can reveal patterns in data, which, when interpreted correctly, lead to facets of understanding not readily apparent. I hope these reflections help you better grasp the tools so that you can build a more informed perspective from apparent "ordinary" data. In so doing, you will find the mystery of evaluation disappears, and you will use the tools to greater advantage in your work.

Along the way to developing an understanding of the tools of evaluation, we will explore several examples of how "discovery learning" can illuminate understanding. I use discovery learning as a primary theme of this book indicating the new insights that can emerge from carefully applied evaluation methods. For example, student achievement has shown itself to be intractable to various remedies. Can our evaluation methods help us see what factors might be productive determinants of student learning? Along similar lines, can educational technology assist the educator and the school district in their work to improve teaching and learning? Beyond education, the world of work is becoming a global environment. Can our evaluation approach help to illuminate the processes that govern a productive work community?

Evaluation methodology can certainly be complex, but it is also very straightforward. At the simplest level of our lives, we recognize correlations, a central evaluation tool. The higher the heat, the more quickly the food cooks; the more our neighbor's dog barks, the more our irritation level grows; the older our cars, the less they are worth; the higher our income, the better we feel about life! But are these seemingly simple intuitions accurate? Aside from the first example, there can be other considerations that might lend themselves to differing interpretations. (Although even the first example is open to new insight if one considers that the altitude can affect how long it takes to cook certain food, and even its quality!) Some neighbors get irritated at dogs barking, but does it depend on other factors like the extent to which we know our neighbor (and the dog), or how "community oriented" our neighbors and ourselves are? Cars get older and lose value, but if they get really old, they gain value. High incomes might lead to greater satisfaction, but not if you live alone—married couples across all income levels generally report they are happy despite how wealthy they are (although it may be desirable to be happily married and wealthy!).

These mundane examples point out that we make observations about our own lives, but we do not necessarily take into account all the factors that might be relevant to a complete understanding. In many cases this is not problematic, although greater insight into our social experience might be helpful to us for greater personal and interpersonal adjustment. Other examples of unexamined correlations are more problematic, however. These are the ones that evaluation methodology can help clarify, and therefore are the focus of this book. Complex programs have far-reaching outcomes and consequences that require specific skills and methods to unravel. The following are the central features in pursuing discovery learning through statistical analysis:

1. Social life presents itself in patterned behavior and thinking.
2. Social behavior and thinking are not determined but predictable.
3. Social research and evaluation studies seek certainty, but must be content with probabilistic conclusions.
4. Evaluation methodology has evolved toward specialization and quantification, both of which contribute to discovery learning.
5. The heart of discovery learning in evaluation is revealed by correlational methods.
6. Many salient insights into social behavior could be more readily discovered by understanding the statistical processes that could illuminate them.
7. Discovery learning insights can better prepare researchers and program evaluators to manage the difficult political and social pressures that often accompany decision-making in the "real world."
8. Evaluation methods cannot eliminate the need for informed judgment.

## INITIAL CONSIDERATIONS

Why does the world need a book that investigates evaluation procedures? The simple answer is that the *concept* of evaluation is largely intuitive, but many people use technical evaluation procedures even when they may not completely understand how to use them to their fullest potential. Few people in a position to need evaluation really understand how powerful it is for illuminating the complexity embedded in programs of interest and thereby avoiding misunderstanding and misdiagnosis. The primary reason for looking at the concept of evaluation is to provide a way to develop better "intuitive insight" into evaluation processes so that we can understand better the social experience we need to examine.

*I feel there is an ongoing need for program evaluation in every social program.* I state it this boldly, and by doing so, I reveal my bias at the outset. Not only is evaluation mandated for certain federal and private grants, but the "spirit" of evaluation points toward improvement of programs and better service delivery. Even where evaluation studies are not mandated, many educational leaders, business managers, social service delivery agencies, and others, have come to recognize that understanding how well you are doing at your tasks will only be positive for introducing changes that can increase efficiency.

Several factors may constrain a program leader from using solid evaluation processes. First, it requires working knowledge of evaluation methods that can illuminate embedded patterns. This is intimidating to some leaders. Second, it can be costly. Evaluation processes may be viewed as unnecessary, or at best, a luxury, especially when they are not mandated. Third, previous, negative, evaluation efforts may be a barrier to using evaluation for subsequent efforts. In many cases the negative attention is due to ineffective methodology or the political dissension stemming from the findings. For these and other reasons many program managers and leaders are reluctant to engage in comprehensive evaluation efforts. This book is devoted to those leaders and to the evaluator who seeks to understand the methods that will improve their attempts at recognizing embedded patterns of meaning. At the very least it will put forward a range of avenues for evaluators to use in their quest to provide understanding and equip leaders with useful information.

## BOOK PLAN

The plan of this book is to look at the concept and practice of evaluation as a way to understand human relationships and to discover "patterned actions" in our personal and social lives. In this discussion I will introduce several statistical techniques that, when used properly, will help the evaluator to understand these patterned actions that impact work, education, and other contexts. Most of the statistical tools we will examine are based in correlation and therefore somewhat intuitive. Beyond correlation, we will spend a good bit of time discussing the technique of multiple regression, which is a technique that allows the evaluator to examine several influences on a particular outcome at the same time.

A book on these topics can be extremely detailed or may only provide basic coverage of key concepts. I want this book to provide depth of understanding on the key areas of regression analysis that emerge from studies in which people work with real data. As such, this book is not a comprehensive mathematical treatment of the procedure, but neither is it a sketchy overview. Understanding what we will cover assumes that the reader has taken an introductory statistics course and understands the basic elements of descriptive and inferential statistics.

This book will not introduce detailed examinations of the various research designs that evaluators use. As we will discuss below, (quasi-) experimental designs and survey approaches are commonly used by evaluators to generate data. While appropriate designs are critical for evaluators, a detailed exposition of research design is beyond the scope of this book. Rather, I will concentrate on the tools the evaluator can use to understand data regardless of the design used to generate the data. The following are some of the tools addressed in subsequent chapters:

- Correlation
- Single predictor regression
- Multiple correlation
- Part and partial correlation
- Detection of extreme scores
- Multiple regression
- Regression with continuous predictors
- Coding of categorical data
- Regression with categorical predictors
- Methods for entering predictors in multiple regression
- Interaction in multiple regression

## REAL EXAMPLES

I have integrated actual databases into the main sections of the book as examples of how to use and interpret correlation and regression results. This decided applied focus to the analyses hopefully will be useful to evaluation researchers as well as to the readers who are new to advanced statistical methods. In all cases I emphasize the point that "discovery learning" can be the result of using statistical processes to understand the unobserved patterns that exist in data.

Before we begin a discussion of correlation and regression methods, I will describe the actual databases that I use to demonstrate the regression procedures. These are not simply collections of data to show how to perform the analyses but examples of actual data that evaluators have used to make policy decisions. I can think of no better way of preparing readers to understand statistical procedures than to provide real data analyses that can cast light on meaningful social issues.

## Using Statistical Programs

Along with extended discussions of the statistical procedures, I describe the use of SPSS® (SPSS® for Windows) as a vital tool for understanding statistical patterns. I show in a practical way how to use the menus in SPSS® to organize and analyze data. SPSS® is a powerful tool that has been used for many years by evaluators and researchers at all levels to extract meaning from data. The calculations and examples in this book require at least a basic familiarity with SPSS®.

## The Evaluator's Journey

Welcome to the journey of developing evaluation skills for discovery. Readers will gain the most from this book if they have a basic understanding of introductory statistical concepts. With this baseline information the book will serve as a guide for evaluation work in many areas of program operation. The book is designed for consultants, evaluators, and other practitioners who are in a position to conduct statistical analyses with data they encounter in their funded (and unfunded) investigations. The methods we explore will assist anyone who seeks to develop a more complete understanding of evaluation processes and how they can illuminate the social world.

Not everyone who reads this book will become an expert in evaluation methods. However, my hope is that all readers will realize that using proved methods appropriately will assist their efforts at discovering crucial patters in evaluation data that will enlighten and improve the programs they are examining.

# 2

# THE ELEMENTS OF EVALUATION

## THE NATURE OF EVALUATION

Evaluation research is known by many names, depending on the purpose of the particular study. It is known as "outcomes assessment" for studies that focus on the end product of program action. "Program evaluation" is another name for studies that have equal focus on the entire structure of an "intervention" or course of action that derives from a set of mission objectives—both the "process of the program" or how it operates and the outcomes. "Impact assessment" and "cost–benefit analyses" are other names for evaluation research, usually in connection to how a program specifically influences the financial and collateral objectives of specific work-related programs. Evaluators can focus on the ongoing "process" of the program, or on the "product" or outcome of the work, or both. "Action research" is incremental problem solving based on reflection and data-driven understanding by a group or individual evaluator.

Regardless of the name, evaluation research usually involves several common elements when it is used in education, social services, organizational psychology, urban planning, and related fields. Program leaders typically are concerned about the progress or outcome of their

*The Program Evaluation Prism: Using Statistical Methods to Discover Patterns,*
by Martin Lee Abbott
Copyright © 2010 John Wiley & Sons, Inc.

programs and seek information about why they have done well, or, more often, why they have not produced the desired results. Evaluation research is simply research that is applied to understand or change a specific program being planned or already in place.

The evaluator is a conveyer of understanding to the sponsor of the research. Through the prism of evaluation methods, the evaluator discerns the goal of the evaluation, creates appropriate designs to capture program efforts, and analyzes resulting data to identify meaningful patterns that may illuminate program behavior.

## EVALUATION CONCERNS

Practitioners need access to sophisticated evaluation models to understand social problems better, but the way in which this understanding is used is often inconsistent and misleading. Careful statistical findings can be used to support a variety of points, depending on the values and interests of the evaluators making and controlling the findings. Nowhere is this more evident than in evaluation research where practitioners and consultants are paid to "solve" specific problems. Are the evaluation findings in these circumstances viewed as a dispassionate record of the relationships under consideration, or are the findings, even unwittingly, used to support the views of those who hired the evaluators? A manager might want to evaluate a certain work practice with an eye toward greater efficiency or productivity, for example, but invalidate findings that run counter to his/her understanding of how the process really works.

In the opening paragraph of this book, I reported on a study I had done on industrial democracy. Let me illustrate the point I just made by reporting on the outcome of that evaluation process. A good number of middle managers were reassigned after I reported my findings! It is probably impossible to disentangle the rationale for these decisions, but one likelihood is that the evaluation findings were used politically when the actual findings were produced with objective statistical methods. That is, evaluation findings may be objectively meaningful and illuminative, but the use of the findings is out of the control of the evaluator. The evaluation findings may have justified those actions, but this illustrates the political and sometimes contested environments within which evaluators work.

The sociopolitics of a setting are therefore likely to nudge the evaluator toward certain conclusions, or patrons may use the findings in a completely unsubstantiated way. People who hire evaluators are typically motivated by the need to show improvement, profit, progress, and so forth, with a primary concern that the evaluation findings will support their understanding of the enterprise. In this setting the evaluator can be as careful as possible but still walk blindly into the misapplication of evaluation results. This does not point blame at the evaluation processes themselves, but rather, it points to the possibility for "objective" knowledge to be "in the mind of the beholder."

Careful evaluators can avoid many of these concerns by adhering to proved statistical and research protocols. But, in conflicted settings, there are severe limitations to what can be found. Usually practitioners have incomplete data, inability to gather meaningful data, are presented with unclear or no program objectives, detect faulty program fidelity, and cannot meet the assumptions for the statistical procedures being used, among other problems. In these situations, is it any wonder that the conclusions may not provide "accurate" and "certain" accounts of human behavior and program action?

Many books can be written (and have been written!) on the shortcomings of evaluation research. I can only note here that evaluators face the same difficulties with quantification that other statistical researchers face. But additional difficulties must be wrestled with that further remove us from making conclusions for which we can have a great deal of confidence. In fact the special difficulty with a great deal of evaluation is that one's conclusions, based on uncertainty and the difficulties noted earlier, might be used for purposes beyond the nature of the research conducted. If evaluation is "purchased," for example, it is open to criticism on many grounds. Here are some further "dangers" that face the evaluation researcher:

- Research performed by untrained researchers
- Inappropriate use of statistical procedures
- Interpretation of results by leaders with little or no statistical knowledge
- Overgeneralization of results
- Use of inferential procedures with inappropriate data sets

Aside from the barriers discussed above, evaluation processes are promising tools that can illuminate recurring social patterns of meaning. It is these specific tools and processes that we will examine in depth in this book.

## EVALUATION STANDARDS

Largely as a result of the evaluation concerns listed above, several standards have emerged that establish appropriate evaluation conduct. In particular, the American Evaluation Association publishes their *Guiding Principles for Evaluators* on their website (http://www.eval. org/GPTraining/GPTrainingOverview.asp). These principles focus not only on the ethics of evaluation but also on the elements of systematic evaluation practice. The principles include systematic inquiry, competence, integrity/honesty, respect for people, and responsibilities for general and public welfare.

Similar ethical guides exist for research in most all disciplinary and professional organizations, like those for the American Sociological Association and the American Psychological Association. However, the guidelines for evaluation practice are particularly relevant to consultants in education and organization study, for example, since their work is particularly vulnerable to the politics of "real life," especially studies that are done outside the published protocols for federal and private foundation grants.

## METHODS USED IN EVALUATION

Evaluators use the same methods that other researchers use to gather important data. In particular, they may use the following approaches in addition to using available existing data:

- Self-designed questionnaires
- Standardized instruments
- Experimental and quasi-experimental design
- Online or phone surveys
- Interviews (structured and unstructured)
- Focus groups
- Observation

Evaluators typically are eclectic in terms of following quantitative or qualitative approaches to understanding data. This book focuses on quantitative methods for understanding data, but it should be clear that solid evaluation work includes any and all means of understanding data, an approach social researchers call "triangulation."

Beyond these methods, however, evaluators are in a position to attend to several dynamics not ordinarily discussed in traditional research books. This is because they must adopt an "executive" perspective on the research focus. It is not enough to analyze data that can help illuminate understanding. The evaluator must also have a conceptual grasp of the context of the data as well as how the program operates.

One important dynamic for evaluators is "program fidelity." This is a term typically applied to considerations of the extent to which a program *actually* operates the way it was intended to operate. If we are not sure the program objectives and methods are operating as designed, how can we attribute the outcomes (or lack of outcomes) to the programs themselves? All the statistical results imaginable, even if done flawlessly, will be unable to establish any connection to the nature of a program if what we are studying is haphazardly implemented or is not operating at all! *Simply observing whether or not a program is operating as designed is often the evaluators best analytical approach.*

## THE EVALUATOR'S TOOLS

Evaluators use a variety of approaches in their work. Many prefer to use experimental or quasi-experimental approaches to generate data, since they feel that these designs are more likely to yield "causal" conclusions. That is, if you can create the experimental conditions that control extraneous influences and isolate the effects of a "treatment" (program, process, etc.) on an outcome variable (worker satisfaction, student achievement, etc.), you can more confidently attribute the change in the outcomes to the treatment. The difficulty, however, is that it is rare for evaluators to be able to create these kinds of conditions. Most of the time we have to do whatever we can to approximate experimental conditions but be content with statistical means of interpreting the resulting data. The classic meaning of control is understood in the context of experimentation. If we were conducting an experiment, we would assume one variable was the cause (independent) and the other was the result (dependent) for the purposes of testing our hypothesis. (However, it might be equally plausible for the "direction of causality" to go the reverse direction.) As an example, does the level of sales motivation among retail workers cause different levels of worker satisfaction, or the reverse? This leads us back to an earlier warning about correlation: Correlation is not causation. There are simply too many other potential "causes" not contained in a correlation relationship for us to say that one variable causes another. Moreover we cannot adequately control an analysis with only two variables.

The alternative to experimental approaches consists of such means as surveys, questionnaires, observation, and other such methods to generate data helpful in understanding a given work or education process. In such cases correlation and regression approaches can be invaluable because they allow the evaluator the ability to "see" how all the variables of interest affect the outcome and one another. The primary limitation, as every evaluator should be aware, is that we cannot assume that correlation results are "causal," since the simple fact that two things are correlated doesn't mean that one thing causes the other to happen. However, we will look at how regression, which is based in correlation, can help to *predict* certain outcomes, and yield a more comprehensive understanding of what might *explain* the relationship among multiple study variables. Used correctly, regression methods can also be used to interpret experimental data. It is therefore the most versatile tool for evaluators in their attempts to understand the dynamics of a given program or process.

## EVALUATION HURDLES

Before we begin our study, it is important to discuss some of the challenges that evaluators face as they grapple with their target programs and the people with whom they work. These are not specific to statistics as a field, but rather concern the conceptual infrastructure within which statistical processes are used. From the earliest social theorists to contemporary public sentiment, the use of statistics has met with serious philosophical objections about using mathematical methods in human inquiry. We will note two of these objections, quantification and specialization, that might be a sticking point for some evaluators as they work in contexts where statistical methods are not fully appreciated.

### Quantification

Quantification is ultimately an act of reducing unobservable realities to measurable arrays of data. This is often nothing more than substituting the objective measure of a behavior for the data referent, as in the case of recording heart rate or the number of widgets produced in a given time span. It is less precise in the case of measuring unobservable things, like a person's opinion about the importance of the widgets produced. The precision is least when we attempt to understand how a person constructs or reflects the meaning of a certain social object, like satisfaction or despair. As evaluators, we are called upon to create data that can help us in discovery learning. However, not everyone involved may share an appreciation for the need for such data.

***Resistance to Quantification*** There is a good bit of resistance to "quantification" and there are many good reasons why quantification can be problematic. Those who favor the "qualitative" approach to evaluation generally, and problem solving specifically, charge that reducing someone's opinion or human characteristics to a number is self-limiting, short sighted, and ultimately misleading. Defining someone's opinion of global warming, for example, may be an artifact of the questions that are asked, some would charge. If you ask people questions, they will give you an answer, but the wording of the question might create categories in the mind of the respondent that were not there previously. Therefore the result of a survey instrument might only be a reflection of the way it was conceptually designed in the first place.

Quantitative research in social sciences has always had its detractors. Among those who would argue the fallibility of social scientific methods are some decision makers who perceive the approach as limiting and perhaps unnecessary. Some groups argue that research as it is typically practiced should place importance on viewing the individual within the contested political and social environment that shapes their consciousness. "Objective" methods are not capable of doing this, and may expose those who are dominated to further subordination.

"Value neutrality" is also criticized as a weakness of quantitative approaches. That is, researchers must curb their own perspectives on whatever social problem they are investigating so that they do not influence or engineer the outcome, even unknowingly. Value neutrality is an impossibility, however. We all see reality through our values. If I view the problem of education as an issue of some social groups underperforming, then I will not be open to understanding other influences that may be at work and operating under my value "radar." Social scientists suggest that one has to adopt a neutral stance to truly see the full nature of a phenomenon. Even though one cannot achieve value neutrality, one can at least admit one's biases and arrange the research process to give all views a fair chance at discovery. The difficulty of doing this is the chief enemy of researchers and evaluators.

***The Nature of Quantification*** The idea that numbers can represent a lot of things not numerical, like relationships, attitudes, and ideas, was a revolution in the human sciences. Deductive theory testing has been used in the physical sciences for many years. Using the same methods allowed social scientists to gather a more "objective" picture of the interior of an individual than that provided by Freudian psychotherapists, for example. A great deal of good may have been done by introducing Freudian concepts into everyday understanding, but these concepts are limited in their ability to identify the objective nature of an individual extricated from his or her social and political environment.

For all their faults, objective methods do provide another "window" into subject matter that is closed to observation. People are not forthcoming with their attitudes, institutional practices are convoluted, and politicized problems are not clearly recognized. Looking at each of these as they are represented in graphs, numbers, and tables allows a researcher the rare opportunity of discovering patterns of behavior that are not otherwise apparent. Few researchers would argue that a person or group studied can be reduced to how they are represented on paper, but many would agree that by making representations, we can see hidden truths.

Here is an example. Look at the following array of numbers:

1, 89, 1, 72, 2, 75, 1, 79, 2, 58, 1, 77, 2, 66, 2, 65, 1, 81, 2, 65.

Simply looking at the numbers tells us little. However, suppose that the 1's and 2's refer to the different methods for teaching a math class to tenth graders: 1's referring to a lesson taught through a "constructivist" method, and 2's referring to a traditional lecture method. The alternating numbers refer to individual students' achievement connected to the teaching method; these numbers, between 58 and 89, represent how well the student understood the content of the lesson as measured by a paper and pencil test (with achievement thus quantified as the results of the test). We can now arrange the data in such a way that a pattern becomes apparent:

| | |
|---|---|
| 1 | 89 |
| 1 | 72 |
| 1 | 79 |
| 1 | 77 |
| 1 | 81 |
| 2 | 75 |
| 2 | 58 |
| 2 | 66 |
| 2 | 65 |
| 2 | 65 |

Brief observation of the two columns of data gives us a bit more information. The 1's are connected to higher scores than the 2's. The 1's scores are in the 70's and 80's whereas the 2's are in the 50's, 60's, and 70's. Thus we could posit that the teaching methods met with different student achievement. Further, if we were to conduct a statistical test (*t* test), we would observe that the two group means are very different from one another. Based on the statistical results, we might conclude that the students (who we assessed to be equal in their understanding of general math before the lesson under study) now are so different they do not belong to the same population of students! The differences after the lessons were so great that they can now be attributed to the different teaching approaches.

Of course, noting the objections to quantification that we mentioned earlier, we would want to comment on the example results listed above. How can we be sure the achievement test adequately represented the math concepts taught? Were all students likely to be equivalent test takers (i.e., paper and pencil tests)? Could different methods (traditional and constructivist) be clearly delineated and differentiated by the students?

Beyond these questions are the concerns of general research protocol. Was the same teacher used for both lessons? In what order were the lessons offered? Were we able to select the students for our different presentations? Were they equivalent learners before the lessons? These and similar questions reflect the issues we have discussed thus far. Quantification is limited; unobservable realities are not easily captured in objective forms. However, we can detect meaningful patterns in data despite the weaknesses of quantification.

***Qualitative Methods***   The statistical enterprise depends on quantification even though it is controversial in many respects. Ironically, qualitative methods can be placed on a continuum of how quantitative they are! At one end of the spectrum are attempts to maintain complete value neutrality through participant observation. In this strategy a person becomes part of the reality that he/she is attempting to study without any of the group under study being aware of the role of the researcher. The result is an attempt to phenomenologically discuss the meaning of the reality studied rather than to present quantified facts amenable for statistical analyses.

The obstacles to this qualitative approach to research are fully discussed in social research texts. I include it here to represent the most "unquantified" of the qualitative approaches. Another method close to this end of the spectrum is the historical or narrative approach to understanding the features of reality of a certain group or setting by examining the "story" that has evolved within the context of the setting and the mind of the participants. Of course, any such methods may gather quantitative observations, but they are most likely used to buttress the qualitative understanding. In the view of some researchers on this end of the spectrum, other researchers who rely on quantification produce results that are an artifact of the way in which they are quantified.

At the other end of the qualitative study spectrum are attempts to quantify observations in straightforward fashion. Usually these involve coding schemes whereby observers record the actual conversations or proceedings of some event under study and then apply a quantitative scheme for discovering the main themes represented in the "data." For example, software is available for identifying certain key words or recurring phrases that might represent enduring patterns embedded in proceedings. Evaluation researchers are familiar with attempts to pose open-ended questions and then training observers to code the meaning of the responses to those questions. Often this results in "percentages of the respondents" who refer to a certain key theme, and so forth. One very creative approach developed by Lofland and Lofland (1995) involved the use of a matrix that identifying "units" and "aspects" of social settings as a guide for coding qualitative data. To this group of qualitative researchers, the meaning of events and human behavior can be better ascertained by using methods that structure the information so that it is more amenable to identifying patterns.

In any case, regardless of the extent to which a method involves quantification, meaning is "extracted" from a phenomenon under study and represented in ways that can be studied apart from the groups or individuals who produced it. Evaluators and practitioners can use these "productions" to identify patterns and digest meaning, but this does not obviate the difficulty of creating certainty using information that is less than certain.

## Specialization

Like all academic disciplines that attempt to explain broad social phenomena, statistics has gravitated toward specialization and away from its early impulses. Early development and use of statistical processes were oriented toward solving specific problems. The narrowing of the inquiries thus fueled the investigation of statistical processes per se and how specific mathematical refinements might yield more accurate results. This resulted in powerful techniques for specific settings and research questions that provide greater confidence in what can be observed from the data.

However, as with other disciplines, has the move toward detailed elaboration left behind the initial motivation of using statistical techniques for solving specific social problems? Stated differently, has the focus on statistical technique outweighed the initial motivations of evaluators who use statistics for illuminating problematic social conditions? The evaluator often encounters attitudes among decision makers that put statistical methods in a negative light because they are overly detailed and "miss the point." The evaluator thus has to walk the fine line between using very detailed procedures and making reasoned interpretations of the results that bear directly on the problem at hand.

## STATISTICAL ISSUES

Evaluators use statistical processes as the means to discover patterns. While it is impossible for anyone to have a thorough grasp of all the statistical tools available, it is nevertheless important to develop a comfort level with using appropriate statistical tools, especially those "tried and true" methods like regression. This is the primary rationale for this book. Before we begin those discussions, however, it is important to look at some of the issues in statistical work that have a direct bearing on the evaluator's work.

### Certainty versus Probability

The nature of the statistical processes that evaluators use is the fact that they are techniques based on probability, not certainty. This is the hallmark of statistics and represents one reason why the field occupies such an important place in research. We can never perfectly predict human action, individually or in groups, no matter how intricate our model or how powerful the procedure. What evaluators hope to accomplish, however, is a more nuanced understanding of the patterns that exist in human behavior, and how those patterns affect the individuals who create and sustain them. This can be a matter of contention for a program sponsor who asks the evaluator to make a concrete decision (certainty) when the data suggest several directions with different likelihoods (probability).

It is important to note, however, that inferential statistical methods rely on a number of processes that attempt to achieve certainty (or as accurate a representation of certainty as possible) that are themselves based on uncertainty. Rules of interpretation have developed to guide the evaluator and scientist toward a view of the findings that is as objective as possible. Statisticians use estimation models and iterative procedures to better approximate the behavior of data, and to better estimate unknown parameters. The "degrees of freedom" used in inferential processes are based on uncertainty and were conceived as adjustments that might achieve more accurate results.

## Statistical Significance

Nowhere is the tension between certainty and probability more apparent than in the language of "statistical significance." This concept has very specific meaning to statisticians and evaluators, and is at the heart of decision-making rules regarding hypothesis testing. Many, if not most, people assume that if something is deemed significant, something "true" has been captured and given the stamp of approval. It is probably the case that the symbols used to convey statistical significance (e.g., $p < .05$) are common references for those who do not understand what the symbols and concept convey.

Even evaluators and researchers may fail to understand the full meaning of the references. Certainly students of statistics have a hard time with the concept until they understand the nature of inferential processes. One misconception is that a smaller $p$ (e.g., $p < .001$) value is *truer*. Recent attempts by statisticians to clarify this concept by focusing on confidence intervals and effect sizes have made some inroads toward a greater understanding of statistical significance.

Students studying correlation now better understand that statistical significance is just an indication of the level of (un)certainty in whether or not a sample value ("statistic") is likely to represent an unknown population ("parameter"). Thus statistical significance implies a question like, Would a sample finding of this magnitude likely come from a population in which there is no relationship between two variables?

## Effect Sizes

Recent emphases on "effect sizes" have helped to place this tension into perspective. If an effect size is the *impact* of the findings, then the question of uncertainty is subordinated to one of meaningfulness. Therefore, regardless of the level of statistical significance, one can ask to what extent the result is *useful* for understanding how one variable practically influences the other. In fact evaluators often refer to effect sizes as "practical significance" over against statistical significance.

To compare these two processes, one might look at a correlation finding. With 100 cases, a correlation of .20 between two variables is *statistically* significant (at the .05 alpha level). However, as we will discuss later in the book, we can create an effect size measure for correlation by simply squaring the correlation. This resulting figure represents how much of the variation in one variable is accounted for by the other variable. In this case the original .20 correlation shows that one variable accounts for only 4% of the variance in the other variable ($.20^2$). This example shows that a finding may be statistically significant but not practically significant. Thus, for example, if the correlation above is a finding for whether socioeconomic status (SES) affects student achievement, I might discover there is a highly statistically significant correlation between the two variables, but the effect size may indicate that, alone, SES has limited utility for explaining much of what affects the variance in student achievement. (If, on the contrary, such a finding resulted in my identifying only 4% of the impact on a new drug to reduce diabetes, I would be extremely happy—and rich!)

## Can We Achieve Certainty?

Of course, more sophisticated statistical procedures like multiple regression exist to help refine an understanding of the relationship between achievement and SES by adding potentially explanatory variables like individual ability, ethnicity, amount of study time in the curriculum, amount of time spent reading, and amount of time spent watching TV or playing video games. But the bottom line is that we are still balancing the tension between certainty and probability. The evaluator can never be completely certain in the attempt to understand human behavior, but a better understanding of the patterns into which we are enmeshed can yield a more informed attempt to remedy or change problematic behavior.

Evaluators who use statistical procedures to design research studies also have developed rules for how best to make the designs more objective and more likely to highlight the contribution of the research variables under consideration. Experimental psychologists, for example, have elaborated the procedures for such investigations to great effect. Much of our knowledge of how people learn can be attributed to such efforts. Ongoing critique about the shortcomings of such processes has also resulted in important observations for social scientists of all stripes.

In any case, there is a trend toward greater precision and objectivity in statistics as a science. However, despite the precision of the evaluator's findings, there is still the likelihood that research sponsors may not use the findings in precise ways. And, on a more prosaic level, we might say that statistical findings are only as good as the training of the one generating the data and using the statistical procedures, regardless of their intentions.

This book is for evaluators who wish to understand the powerful tools of research that can assist with their program evaluation projects. I am also motivated to provide a deeper understanding of statistical processes that are thought to be intimidating but that can be helpful in producing meaning. This last hope is occasioned by the recent books that use statistical processes (especially regression) but do not progress beyond a topical discussion of the insights that can emerge from using the processes.

## Dispelling the Mystique of Statistics

The first task of the evaluator is to understand the swirling of concepts and misperceptions about statistics prevalent in common understanding and thereby to gain a statistical perspective on how to use these powerful tools to solve problems. Statistical procedures are tools brought to bear on the sometimes tough problems evaluators encounter as they seek approaches to understand the myriad data sources of targeted programs.

Statistical procedures intimidate some people, especially those with a fear of math. While this seems to be a generic feature of our education system, this fear does not have to prevent anyone from appreciating and using statistical procedures. It is true, however, that human beings cannot learn in a context of high anxiety. As support for this notion, I often tell my students to try thinking about the use of irony in Dostoyevsky's *Crime and Punishment* (1866) while under emotional distress. Most would not be able to overcome the affect that threatens cognitive understanding. In my experience, this is how some evaluators feel about using statistical methods. Using more complex procedures may be avoided by using procedures that cause lesser anxiety about how to interpret available data.

This book is designed to help evaluators overcome these fears through an understanding of statistical methods, and how statistical procedures can be effectively used. In particular, it is designed to help evaluators who have some introductory understanding of statistical processes, possibly in college courses they have taken. There are many advanced statistical techniques available to the evaluator once they have mastered the introductory concepts. We will focus on correlation and regression methods in this book because of their widespread use (and misuse) and because of their usefulness in program evaluation.

In this book we will also introduce several methods to help dispel some of the mystique of statistics. First, we will use actual databases so that you can get an "over-the-shoulder" view of how to analyze real-world data. Second, we will use SPSS® extensively to show how to manage data and how to interpret findings in a step-by-step fashion. Third, we will focus our study only on correlation and regression methods so that you can have a comprehensive grasp of the procedures before you launch an investigation of related statistical methods.

## Research Literacy

Often, even well-intentioned people can, through lack of knowledge about appropriate statistical processes, show something to be true that is not, and vice versa. Further they can create policies that have a negative impact because the basis for deriving the policy is heavily flawed. Statistics are tools that can be misused and influenced by the value perspective of the wielder. Most evaluators have stories about how unfounded assumptions or prejudices have driven decisions and may have even generated unfair outcomes. These are all reasons to become aware of how statistical processes should be used as well as the limits of our knowledge.

One of the databases we will use in our discussion of statistical procedures is taken from student achievement results in the state of Washington. As you may know, there is a continuing public debate over federally mandated testing and test results in our society. I wish to use current data to show how statistical procedures can help to illuminate current social questions and perspectives. The education system is a politicized and often contested landscape that is constantly changing. Educational systems represent the hopes and frustrations of parents, administrators, and a variety of activist groups, in addition to the students who "live it out." We hear regularly about what works best to assist students in learning, and we consider the stream of mandates for more efficient and productive school systems. Increasingly, approaches to effectiveness derived from the business world are being applied to educational institutions with mixed results.

Our ability to generate and process information is growing exponentially. More than ever, teachers, principals, and evaluators need to be in command of what is being studied and reported by those who produce research articles, monographs, evaluation reports, and the like. More and more information is available, and there are many attempts to assert "truth" about how it is all to be done best. Unfortunately, these attempts are not always made with appropriate attention to the statistical methods that have been accepted as reasonable guides to understanding and acquiring knowledge.

In short, we need to become "research literate" if we are to develop the ability to sift out the appropriate from the inappropriate attempts to establish what comes to be accepted as "truth." This is an often intimidating prospect. While the processes are complex and can be bewildering, a basic understanding of the way research works will go very far to help decipher what is being said, whether or not the conclusions are warranted, and how further steps can be taken to shed light on the initial questions.

## THE DISCOVERY QUESTIONS

This book is about using statistics as a way for evaluators to discover hidden patterns of meaning. While we will spend a good bit of time discussing the details of the methods, we also want to use the methods to help us discover potential answers to some intriguing questions.

I will use examples from several databases in this book as a way to illustrate discovery learning. These examples come from my evaluation work in education and organizations and illustrate how evaluation research can yield new insights. The data analyses procedures discussed in this book can be used in a wide variety of areas. The problem areas I chose represent current evaluation concerns in contemporary society.

### School Characteristics and Student Learning

In recent years educators have pointed to the "achievement gap" among students of different ethnic backgrounds, suggesting that (especially) nonwhite students suffer particular disadvantages that affect their achievement rates. This notion is recognized in the No Child Left Behind (NCLB) legislation that targets remedial efforts for those schools that do not evidence improvement in achievement among students of different ethnic groups and other subgroups. Apart from funding efforts, state, district, and school leaders have attempted a variety of programs to bolster greater participation and success among all these subgroups. School district leaders have become concerned as they attempt to respond to the federal mandates. What can these leaders do to eliminate the gaps targeted by NCLB?

In trying to understand these relationships by looking at the achievement results in Washington schools, we recognize that designations are complex realities, and simple correlations with achievement may not fully represent the many facets of these realities. My research has generally focused on possible determinants for academic achievement among some of these subgroups of students. What could account for the different levels of achievement evidenced among school children at all levels: parent involvement, student aptitude, ethnicity, socioeconomic status, teaching methods, school or class size, classroom technology, different curricula, influences from home, and peer culture?

In my evaluation work, both individually and as part of larger evaluation efforts, it is fair to say that there is no definitive answer to this question. Probably all of the factors noted above influence student academic achievement. Perhaps there is no "super" factor that represents the greatest influence. However, research does indicate that some of the assumptions prevalent in the public media about student achievement may not be completely accurate. We will use some of the evaluation data to examine some of these areas in this book.

## Worker Participation

Several years ago it was fashionable to encourage businesses to adopt "industrial democracy" as an attempt to engender loyalty to the company and to create a workplace community. This was fashioned largely from the Japanese approach that Ouchi (1981) called "theory Z," in contrast with the American approach that stressed "theory Y" management. The American strand was based on social science principles that suggested older models of management ("theory X") were overly structured, harsh, and placed no emphasis on the worker as an individual but rather only as a part of the organizational system that could be tinkered with to create better productivity. Latent reasons for adopting a more worker-centered philosophy included greater worker productivity that might ensue from workers taking greater ownership in the organization and worker satisfaction, which might alleviate stress, absenteeism, and vandalism.

The newer Japanese management models were reflected in very successful companies (especially in the automobile industry) that grew by engendering loyalty and sense of place. Worker commitment would be a natural by-product of participating in a company that was interpreted by the worker not as just a place to earn an income, but rather a place where they belonged—where they were part of a vital team.

Although these management principles are discussed in different terms in contemporary studies, the legacy of the shift in management focus can be recognized in recent companies that are entering the globalization phenomenon that Friedman (2005) discussed in *The World Is Flat*. What appears to have impacted business operations the most involves forces "outside" the organization itself. That is, changes in technology, communication, and the explosion of knowledge have enabled business leaders and workers to form international communities focused on global markets and on accomplishing tasks that used to be encapsulated and inward-focused among individual businesses. What effect this is having on workers and on the nature of industrial democracy is yet to be determined. However, it is fair to say at this point that workers understand in a poignant way that their work now takes place in a much larger arena and that they will have to look differently at their jobs.

My evaluation studies in industrial democracy help illustrate the principle that what appears to be a causal relationship between worker commitment and company outcomes may be more complex. Simply instituting a worker participation system in a business may not necessarily lead directly to worker outcomes like success or satisfaction. The truth of the matter is that other things might explain why workers gain satisfaction. The workers' desire for participation or acceptance of the opportunity to participate would seem to be a precondition for industrial democracy to work effectively. Also participation might be a function of a workers predisposition or basic style of interaction (introversion–extraversion, etc.). Some workers may accept the opportunity and use it as intended by the organization designers, while others may not. In other words, there may be several other variables that make the linkage between worker participation and worker satisfaction, for example, more complex and not causal.

When I asked workers in a manufacturing plant whether they wanted to participate in company programs where workers could take part in decision-making, not everyone agreed that it was a positive option. Several workers desired participation and took advantage of the opportunity to affect decisions that directly impacted their work. However, a sizable proportion of them did not, and statistical techniques like the ones we are discussing helped point out that things are not always as they seem.

This dynamic in worker participation is similar to educational questions like whether school size affects learning, for example. Management initiatives in decision-making might lead to participation just as smaller schools might lead to better student learning. However, these changes may not produce the anticipated results. Moreover, even if the desired results are achieved, they really may depend on a host of factors that might not be immediately apparent, but are exerting strong influences nevertheless.

## The Impact of Technology on the Classroom

Over the years I have conducted several evaluation studies using correlation among variables from large databases. One such study involved a nationwide online survey focusing on the impact of classroom technology on school and classroom outcomes. The data came from the TAGLIT (Taking a Good Look at Instructional Technology) project[1] and involved survey responses from elementary and secondary teachers and students and school administrators regarding the potential impacts of instructional technology on a variety of school and classroom outcomes. The instruments were originally developed by educators in the University of North Carolina and administered to develop and understand the nature of school leadership. It is currently housed and operated by T.E.S.T. Inc.

The TAGLIT instrument is actually a series of instruments administered to several different groups within schools in the attempt to understand the extent of technological literacy, and the impact technology might have on different student, teacher, and school outcomes. In 2003, assessments of these different groups focused on several aspects of technology in the classroom, including the following:

- The technology skill levels of teachers and students in elementary and secondary schools
- How often technology applications were used in learning
- What access to technology did teachers have
- The extent of professional development of teachers
- School expenditures for technology (including support personnel)

Several of the items were common to both student and teacher groups, which permitted direct comparison of the responses. One such area was a series of questions that focused on the extent to which certain technological applications were being used in the classroom

---

[1] The author acknowledges the kind approval of T.E.S.T., Inc., the owner and manager of the earlier TAGLIT data for the use of TAGLIT databases for this book. (http://www.testkids.com/taglit/)

(e.g., word processing, spreadsheets, WWW, email, or presentation software). Because these same questions were asked of the different respondent groups, the researcher could compare the extent to which the individual students and teachers reported that they used the various applications. The instruments asked not only the perceived technological skill level of the respondents with the applications but also whether, or to what extent, the applications were used in classroom learning.

Another series of items assessed the understanding of, and use of, the various technology applications in classroom learning by teachers. A similar series of items were asked of students so that comparisons could be made between student and teacher perceptions of the role of technology in student learning.

A very important part of the TAGLIT tests was the gathering of factual information from school leaders as to costs and expenditures of technology so that these data could be used to compare with teacher reports of technology availability and support for classroom use. It is often the case that evaluation projects rely solely on the opinions of survey respondents for information about the nature and viability of the program under consideration. However, respondent's perceptions alone are not always reliable, so it is always desirable to bring additional information to bear from different sources to create an objective frame of reference to study results. In the case of TAGLIT, the school leaders' responses provided this frame of reference.

The TAGLIT data provided a rich source of information about the understanding and use of technology in teaching and learning across the country. Year after year the data indicated similar results with respect to teachers' and students' technology skill and how this affected the classroom learning process. There is a connection between the teacher's technology skill and how learning takes place among students. We will explore some of these connections in our discussion below of correlation and regression methods.

## Classroom Observation Data

An important source of data in educational research is observation in the classroom. Most often education researchers are forced to rely on secondary data sets at the school level to understand the determinants of individual student achievement. Individual level data collected under controlled conditions is rare. Typically researchers have a limited sample of schools from which to gather individual data, and therefore have difficulty moving beyond case study description in their findings and recommendations.

One exception to this type of educational research is observational data generated at the classroom level. Later in the book I will present data to show how observation data can be used to provide insight into the nature of academic achievement. The data in the dataset we will use are unusual because they are so extensive. The BERC Group, Inc. conducted thousands of classroom observations over the last few years in hundreds of districts in order to understand the nature of teaching and learning in various subjects at elementary and middle schools in the five essential components of the STAR Classroom Observation Protocol™. The STAR Classroom Observation Protocol™ is a research-based instrument designed to measure the degree to which Powerful Teaching and Learning™ is present during a classroom observation.[2] We will use these data and connect them to school-level data.

These and other related efforts to understand what happens in the classroom and its impact on achievement are the focus of several recent studies. One study I published with my colleagues (Abbott et al., 2010) focuses on the nature of math teaching and the extent to which meaningful educational outcomes depend on support for teachers and developing teacher knowledge in math.

## DISCOVERY LEARNING

### Terms and Concepts

**Effect size**    A measure of the impact of findings, as the extent to which one variable practically influences the other (also known as "practical significance").

**Evaluation research**    Research methods applied to understand or change a specific program being planned or already in place. Evaluation can have many names, each emphasizing a different feature:

- Outcomes assessment
- Program evaluation
- Impact assessment
- Cost–benefit analysis
- Action research

---

[2]These data are used with the permission of The BERC Group, Inc.

**Evaluation standards**   Agreed-upon principles that focus on the ethics of evaluation and the elements of systematic evaluation practice.

**Program fidelity**   The extent to which a program actually operates the way it was intended to operate.

**Qualitative methods**   Using information to understand the nature of an object or situation without reducing it to a numerical referent.

**Quantification**   The act of reducing unobservable realities to measurable arrays of data.

**Statistical significance**   An indication of the level of (un)certainty in whether or not a sample value ("statistic") is likely to represent an unknown population ("parameter").

# 3

# USING SPSS®

In a book such as this, it is important to understand the nature and uses of a statistical program like SPSS®. There are several statistical software packages available for manipulation and analysis of data, but in my long experience SPSS® is the most versatile and responsive program. Because it is designed for a great many statistical procedures, we cannot hope to cover the full range of tools within SPSS® in our treatment. We will cover, in as much depth as possible, the general procedures of SPSS®, especially those that provide analyses for correlation and regression. The wide range of SPSS® products is available for purchase online (http://spss.com).

The calculations and examples in this book require a basic familiarity with SPSS®. Generations of social science students and evaluators have used this statistical software, making it somewhat a standard in the field of statistical analyses. In the following sections, I will make use of SPSS® output with actual data in order to explore the power of statistics for discovery. I will illustrate the SPSS® menus so it is easier for you to negotiate the program. The best preparation for the procedures we discuss, and for research in general, is to become acquainted with the SPSS® data managing functions and menus. Once you have a familiarity with these processes, you can use the analysis menus to help you with more complex methods.

*The Program Evaluation Prism: Using Statistical Methods to Discover Patterns,*
by Martin Lee Abbott
Copyright © 2010 John Wiley & Sons, Inc.

There are several texts that use SPSS® as a teaching tool for important statistical procedures. If you wish to explore all the features of SPSS® in more detail, you might seek out a source such as Green and Salkind (2008).

## GENERAL FEATURES

Generally, SPSS® is a large spreadsheet that allows the evaluator to enter, manipulate, and analyze data of various types through a series of drop-down menus. The screen in Figure 3.1 shows the opening page where data can be entered. The tab on the bottom left of the screen identifies this as the "Data View" so that the evaluator can see the data as they are entered.

A second view is available when first opening the program as indicated by the "Variable View" also located in the bottom left of the screen. As shown in Figure 3.2, the Variable View allows the evaluator to see how variables are named, the width of the column, number of decimals, variable labels, any values assigned to data, missing number identifiers, and so forth. The information can be edited within the cells or by the use of the drop-down menus, especially the "Data" menu at the top of the screen. One of the important features on this page is the

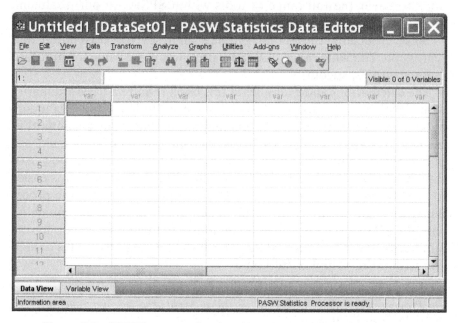

**Figure 3.1** SPSS® screen showing data page and drop-down menus.

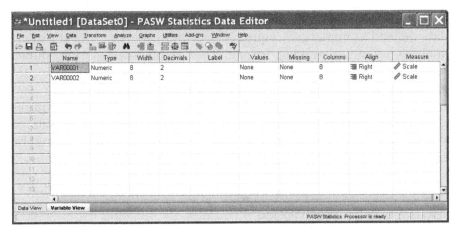

**Figure 3.2**    SPSS® screen showing the variable view and variable attributes.

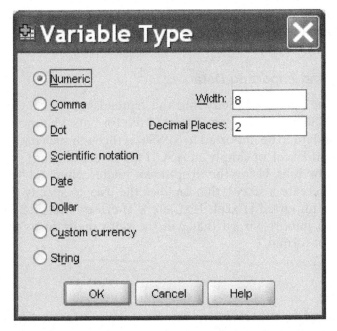

**Figure 3.3**    SPSS® screen showing submenu for specifying the type of variable used in the data field.

"Type" column, which allows the evaluator to specify whether the variable is a number, a letter, or some other form.

Figure 3.3 shows the submenu available if you click on the right side of the Type column in the Variable View. This menu allows the evaluator to specify the nature of the data. For most analyses, having the data defined as numeric is required, since most (parametric) statistical analyses require a number format. The "String" designation, shown below at the bottom of the choices, allows the evaluator to enter data as letters and words, such as quotes from research subjects and names of subject groups. If you use a statistical procedure that requires numbers, make sure the variable is entered as a "numeric" variable, or you will get an error message and your requested procedure will not be executed.

## MANAGEMENT FUNCTIONS

In this "crash course" of SPSS®, I will cover the essential functions that allow the evaluator to get started right away with the analyses. Before a statistical procedure is created, however, it is important to understand how to manage the data file.

### Reading and Importing Data

Data can be entered directly into the "spreadsheet" or it can be read by the SPSS® program from different file formats. The most common format for data to be imported to SPSS® is through such data programs as Microsoft Excel, or simply an ASCII file where data are entered and separated by tabs. Using the drop-down menu command "File-Open-Data" will create a screen that enables the user to specify the type of data to be imported (Excel, Text, etc.). The user will then be guided through an import wizard that will translate the data to the SPSS® spreadsheet format.

### Sort

It is often quite important to view a variable organized by size, and so on. You can run a statistical procedure, but it is a good idea to check the "position" of the data in the database to make sure the data are treated as you would expect. In order to create this organization, you can "sort" the data entries of a variable by choosing sort on the drop-down menu for "Data" as shown in Figure 3.4. You can specify "Sort Cases" and get the following data screen. In the example below, if we place "MathPercentMetStandard" in the "Sort by:"

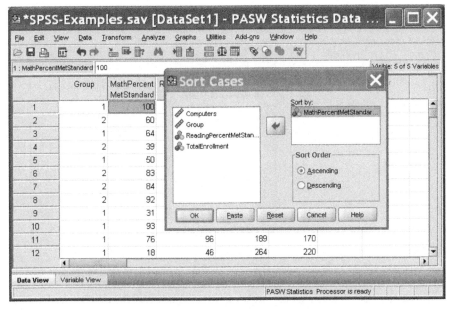

**Figure 3.4**    SPSS® screen showing the "Sort Cases" function.

window, it will arrange the data either in "Ascending" or "Descending" order.

## Split File

A very useful command that we will use extensively in the chapters below is "Split File," which allows the user to arrange output specifically for the different values of a variable. For example, in Figure 3.5, I have created submenus for Split File by choosing the Data drop-down menu, and then selecting Split File. As the screen shows, I have selected the option "Organize output by groups" and then clicked on the variable "Group" in the database. When I create this action, I am issuing the command to create (in this case) two separate analyses for whatever statistical procedure I call for next since there are two values for the Group variable ("1" and "2"). I can then call for a correlation analysis, for example, and the output file will show two correlations, one for group 1 and one for group 2.

It is very important to remember to go back to this option once your overall analysis is over. SPSS® will continue to provide split file analyses until you "turn it off" by selecting the first option, "Analyze all cases, do not create groups" at the top of the option list.

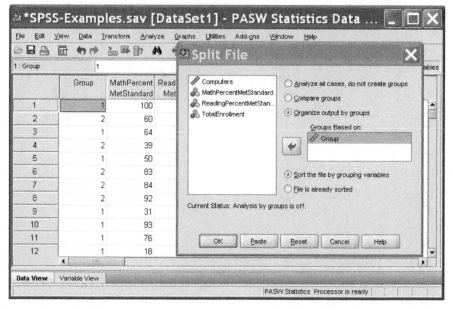

**Figure 3.5** SPSS® screen showing the "Split File" function and submenus.

## Transform/Compute (Creating Indexes)

One of the more useful management operations is the Compute function, which allows the user to create new variables. At the main menu, I can access this function by selecting the "Transform" and then "Compute Variable" option. This will result in a dialogue box like the one shown in Figure 3.6. In this example I am creating a new variable ("compstud") by dividing the current "Computers" variable by the "TotalEnrollment" variable. If the database consisted of school-level information, this new variable might represent the proportion of computers per student at each school. As you can see from the screen, you can use the keypad in the center of the dialogue box for entering arithmetic operators, or you can simply type in the information in the "Numeric Expression:" window at the top. You will also note that there are several "Function group:" options at the right in a separate window. These are operations grouped according to type. Scrolling to the bottom allows the user to specify "statistical functions" like means and standard deviations. You can select whichever operation you need, and enter it into the Numeric Expression window by clicking on the up arrow next to the Function Group window.

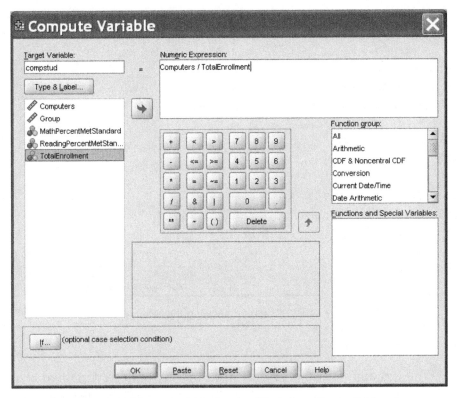

**Figure 3.6**   SPSS® screen showing the "Transform Compute" functions.

## Merge

The "Merge" function is one of the most useful, but the most misunderstood functions in SPSS®. I have yet to see any accurate treatment of the appropriate steps for this procedure in any resource book. I will attempt to provide a brief introduction to the procedure here because it is so important, but experience is the best way to master the technique. I recommend that you create two sample files and experiment with how to use it.

The merge function allows you to add information to one file from another using a common identifier on which the procedure is "keyed." For example, suppose that you are working with two separate school-level data files and you need to create one file that contains variables located on the separate data files. Perhaps one file has a school ID and the "computers per student" we created in the last example, while a second file has a school ID (the same values as the other file) and achievement results.

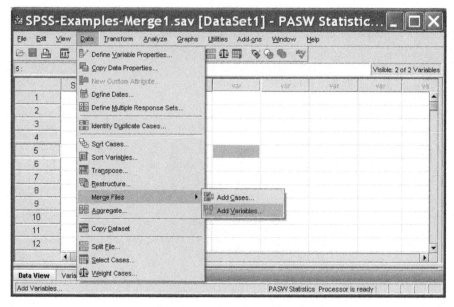

**Figure 3.7** SPSS® screen showing the "Merge" options.

You can approach this several ways, but my preferred method is to choose one file as the "master" to which the separate information is brought. After the transfer you can save this file separately as the master file, since it will contain both sets of information. SPSS® allows you to specify which information to take to which file on a separate dialogue box.

The first step is to make sure both SchoolID variables are sorted (Ascending) and saved the same way. This variable is the one on which the sort is keyed, and the merge cannot take place if the variables are sorted differently within the different files, that is, if there are duplicate numbers, missing numbers, and the like.

Second, in the file you identify as the master file, choose "Data–Merge–Add Variables" as shown in the dialogue box in Figure 3.7. This will allow you to move entire variables from one file to another. The other option "Add Cases" allows you to append all the *cases* in one file to the cases in the other file, a completely different function.

When you ask to "add variables," the dialogue box shown in Figure 3.8 appears, which enables you to choose the data file from which you wish to move the desired variable. The next dialogue box that appears is shown in Figure 3.9, in which you can specify which

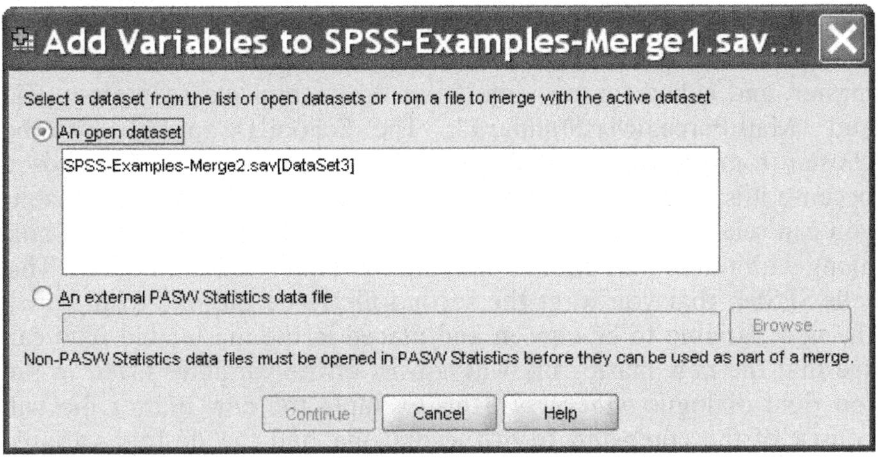

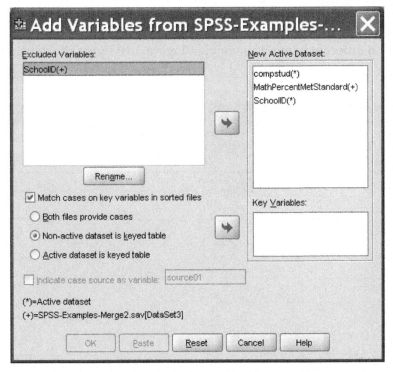

**Figure 3.8**    SPSS® screen showing submenus for the "Merge" function.

**Figure 3.9**    SPSS® screen in which users can identify the "key variables" on which to base the merge.

variable is the "key variable" on which to base the merge. In the current example I have used the first file (containing ID and "compstud" as the master, and called for a merge from the second file (containing ID and "MathPercentMetStandard"). The SchoolID variable can be chosen from the top left dialogue box. It is the only variable chosen because it is found in both files. After selecting the SchoolID variable, you can select the "Match cases on key variables in sorted files" box, along with the middle choice "Non-active dataset is keyed table." This tells SPSS® that you want the second file to be the one from which the new variable to be chosen and placed in the master file. You can see that the new master file will consist of the variables listed in the top right dialogue window. In this example the new master file will consist of the compstud from the first file, and the desired variable ("Math …").

The next step is very important. When you click on the bottom arrow, you inform SPSS® that the SchoolID variable is the keyed variable. When you place SchoolID in this window, it removes it from the list of variables in the "New Active Dataset" window, since it is contained in both files. Once you perform this action, you can choose "OK," and the desired variables will be added to the master file, which you can save under a different name. The panel in Figure 3.10 shows the new master

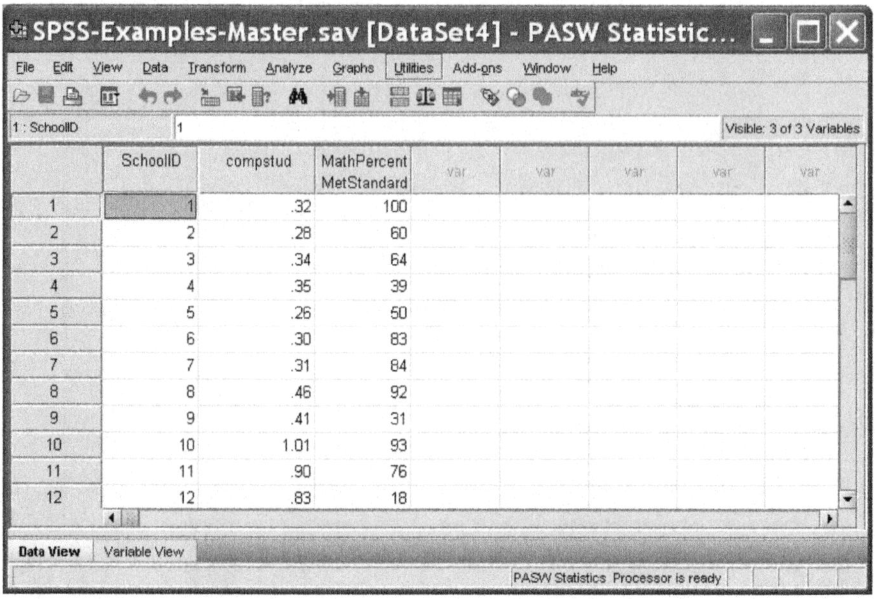

**Figure 3.10** SPSS® screen showing merged file.

file with the desired variables (compstud and MathPercentMetStandard) keyed to the same SchoolID number.

## ANALYSIS FUNCTIONS

Over the course of our study in this book, we will have extensive practice at creating statistical procedures. For the most part these will focus on correlation and regression procedures. All of these are accessible through the opening "Analyze" drop-down menu. The screen in Figure 3.11 shows the contents of the Analyze menu, and that I have specified the "Correlate" procedure. This creates a submenu for the procedure in which I can specify a "Bivariate" (two-variable) correlation, a "Partial" correlation, and so forth. We will review these submenus extensively as we cover the procedures in subsequent chapters.

## GRAPHING FUNCTIONS

SPSS® has many graphing options for displaying data. We cannot hope to cover all these options in this book, but I want to introduce the

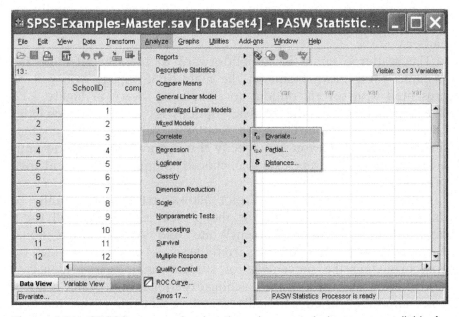

**Figure 3.11** SPSS® screen showing the primary analysis menus available for specific statistical procedures.

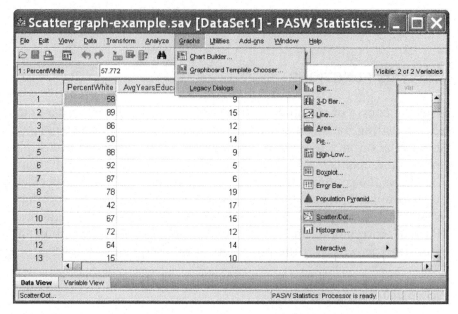

**Figure 3.12**   SPSS® screen showing available graphing options.

scatter diagram option, since these are helpful for visualizing correlation relationships. For information about the graphing capability of SPSS®, consult the Green and Salkind (2008) source on SPSS® I cited earlier.

The scatter diagram is straightforward to use. By selecting the "Graphs" drop-down menu on the main SPSS® page, you will be presented with several options for graphs as shown in Figure 3.12. "Legacy dialogs" are the "classic" options for graphs and includes "Scatter/Dot ..." graphs. Choosing "Scatter ..." will bring you to another dialogue box where you can specify the $X$ and $Y$ variables in your analysis. For the present example, choose "Simple Scatter" and then "Define."

Calling for a Simple Scatterplot allows you the opportunity to fine-tune a scatter diagram. For the present example, I call for a scatter diagram between "PercentWhite" and "AvgYearsEducational Experience." These are variables from the school database in which we are correlating the percentage of students at a school identified as Caucasian, and the average teaching experience at the same schools. One supposition in educational research is that the newest teachers are often assigned to the poorest schools with the greatest numbers of nonwhite students, and such practice could present a structural

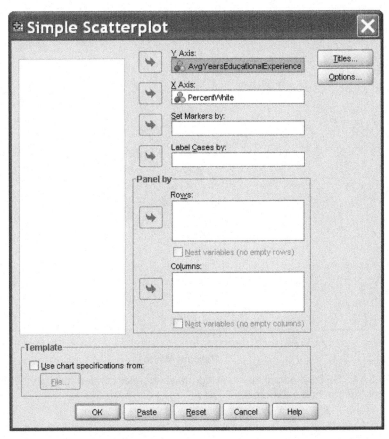

**Figure 3.13**   SPSS® screen showing a submenu for specifying the parameters for creating a scatterplot.

condition for new teachers failing in their initial teaching experience. The scatter diagram will provide some visual clues about whether this supposition is accurate.

The graph specification in Figure 3.13 calls for a scatter diagram with the percentage of white students at the sample schools as the predictor variable, and average teaching experience in the schools as the outcome variable. Selecting "OK" to the scatter diagram dialogue box will result in a scatter graph being generated in the SPSS® output file. The resulting graph is shown in Figure 3.14. Each school is represented as a dot on the graph. In the next chapter where we discuss correlation, we will review these graphs extensively. At this point I present this information as a way of preparing you to create visual referents from which to discover patterns in data.

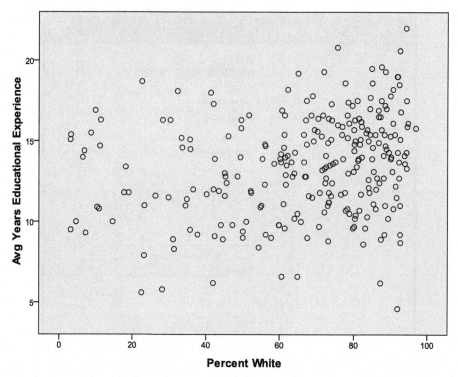

**Figure 3.14**   SPSS® output file showing the scatterplot created through the use of the graph menus and submenus.

An additional step for the graphs that we will explore in the next chapter is to add a "best-fitting" line, which is a line that is passed through the data in such a way that the distance from each dot to the line is at a minimum. (This is the "regression line" that we will discuss in the chapter 5) To add this line, double click anywhere on the graph that is displayed in the output file.

Double clicking on the graph will create a "Chart Editor" dialogue box in which you can select the fifth button from the left edge of the second row of menu icons. Holding the cursor above this button will read "Add Fit Line at Total." You can see the Chart Editor box in Figure 3.15. Selecting this button will result in a line of best fit being added to the scatter diagram. This new scatter diagram is shown in Figure 3.16.

We will have more to say about these kinds of graphs in the next few chapters. At this point I wanted you to develop the ability to create the graphs and use them in interpretation. As you can see from the graph

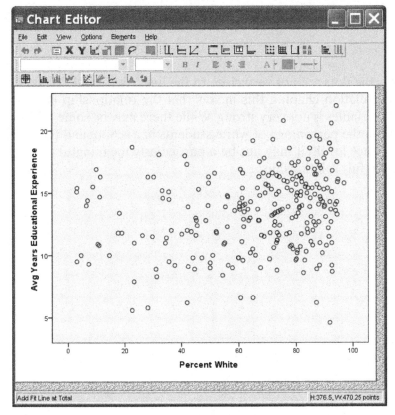

**Figure 3.15** SPSS® screen showing the scatterplot generated in the output file that can be edited by double clicking in the graph area.

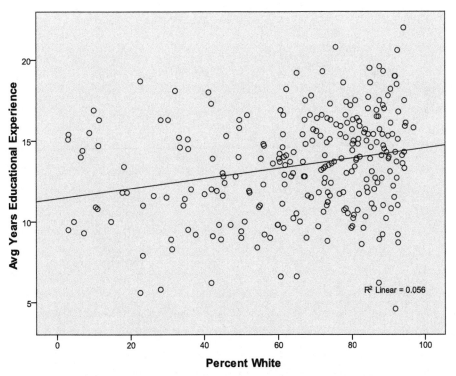

**Figure 3.16** SPSS® screen showing scatterplot with regression line.

above, the dots are not very close to the line. As we will discuss next in the correlation chapter, this means that the relationship between the study variables is not very strong. While there may be some relationship between the percentage of white students in a school and the teachers' experience levels, it may not be a particularly meaningful (statistically meaningful) relationship.

# 4

# CORRELATION

## THE NATURE OF CORRELATION

Correlation has an intuitive appeal; people seem to understand that changes in one thing are related to changes in something else. Thus, for example, we may observe that students who get the highest reading achievement test scores are also the ones who read the most. Or, stated another way, as the achievement scores change in the upward direction, so does the amount of time spent reading, and vice versa. We might represent this relationship graphically as in Figure 4.1.

By graphing each student's scores at the same time, we get a "scatterplot" that indicates how they all relate to each other. This example is a "positive" correlation or association, since as students increase their values on one variable, they also increase on the other. The data in Figure 4.1 are hypothetical, but they do reflect some of the studies we have done on determinants of reading achievement. The data show that time spent reading per week (the $x$ axis) is related to reading achievement scores ($y$ axis). The bottom panel in the diagram shows the pattern of the scores as laying close together in an upward direction. This relationship expresses a "positive correlation" in that increases in the values of both variables occur together. A "negative" correlation could also occur, as shown in Figure 4.2.

*The Program Evaluation Prism: Using Statistical Methods to Discover Patterns,*
by Martin Lee Abbott
Copyright © 2010 John Wiley & Sons, Inc.

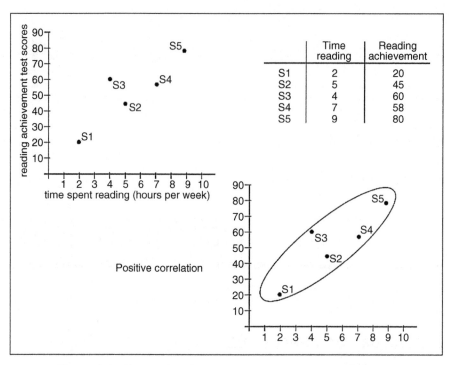

Reading achievement data table:

| | Time reading | Reading achievement |
|----|----|----|
| S1 | 2 | 20 |
| S2 | 5 | 45 |
| S3 | 4 | 60 |
| S4 | 7 | 58 |
| S5 | 9 | 80 |

**Figure 4.1**   Scatterplot of data on two variables simultaneously.

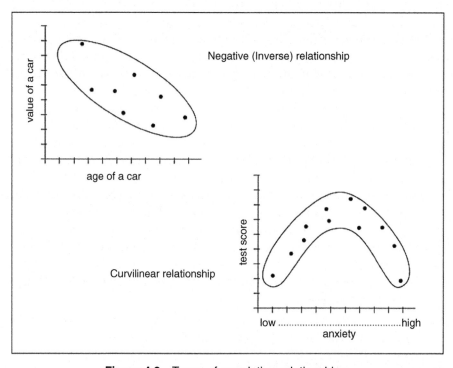

**Figure 4.2**   Types of correlation relationships.

In the top panel of Figure 4.2, the graph shows that as the age of a car increases, its value decreases; thus a "negative" relationship is characterized by a downward trend in the scatter of scores. Actually "negative" does not have a value connotation in correlation analysis; it does not mean "bad." It simply refers to the "direction" of the relationship. A negative correlation is also known as an "inverse" relationship.

Figure 4.2 also shows another possibility for correlation results in the bottom panel, a curvilinear relationship. We will not discuss curvilinear relationships extensively in this book, but rather will examine "linear" relationships where the variables relate to one another along a straight trajectory. In the bottom panel of Figure 4.2, the data for anxiety and learning show a curvilinear relationship. The panel shows that low to medium levels of anxiety are positively related to performance (test scores in this example), but medium to high levels of anxiety are inversely related to performance. This example is similar to classic studies of anxiety and performance in the social science literature.

You probably recognized the fact that the (negative) relationship shown in the top panel could really have become a curvilinear relationship if we had extended the x axis a bit further. Older cars may be worth a lot!

There can also be a "zero correlation" between variables if the values of one variable do not change as the other variable changes. This would be represented by a graph of scores that have no pattern to them—upward, downward, or curve-like. Figure 4.3 shows how a zero correlation would appear on a graph.

## PREDICTION

Correlations are valuable for several reasons. They express visually and numerically what the relationship may be between the evaluator's study variables. They further express the strength of those relationships. Beyond these is the fact that correlation procedures are basic to the process of regression, which allows the evaluator to predict outcomes more accurately given their relationship to study inputs. That is the importance of recognizing whether or not two variables are strongly related (correlated) to one another. Take the example in Figure 4.1. If we know a relationship exists between hours reading and achievement test scores, we could use this information in a regression analysis to predict that a new student in our class will likely perform in the same fashion, and we can predict her achievement score from knowing how many hours she reads each night. Or, following the data from the top

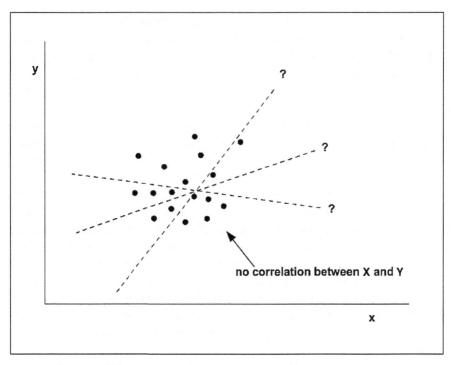

**Figure 4.3**   Types of correlation relationships.

panel of Figure 4.2, we could predict the value of our car if we simply knew the age of the car.

Of course, these are simplistic examples. More realistic examples would be based on larger, more carefully controlled data sets containing many more study variables. However, the examples underscore the fact that correlation is extremely valuable as a prediction device. In making this statement, we should note that a significant correlation allows us to predict the value of one variable when we know the value of the other; it does not allow a conclusion that one variable "causes" the other.

## CORRELATION IS NOT CAUSATION

One problem with discovering significant correlations is the temptation to assume that the variables are *causally* related. Thus we might be tempted to assume that our correlation means that time reading *causes* achievement test score increases. It doesn't, necessarily. There are a great many things that are related to reading achievement test scores, and some of them may be better explanations for increases in achievement scores, even though time spent reading is strongly related. If you were to list other *potential* causes, you could probably produce a long list, for example, the nature of *how* the student reads, the student's ability, school curriculum, home environment, interest level, parents' education level, and so on.

Let me say a word about the terms we use in this book for the variables in correlation and regression analyses. In some discussions, variables can be termed "independent" or "dependent," according to the nature of the study. These terms seem to be widely accepted by evaluators. However, these labels are not accurate for every context involving two variables. They might be appropriate in an experimental context if we were comparing experimental teaching methods with traditional teaching methods on student achievement, for example. The labels would not be as appropriate if we were only interested in describing the ability of one variable (e.g., amount of daily homework) to predict the value of another (e.g., test score). In the latter case we might suggest that the predictor be called just that, "predictor," and the other variable the "outcome" or, as some would suggest, the "criterion" variable. We will keep this convention in our following discussions.

## PEARSON'S *R*

If we want to measure precisely the relationship between variables, we can use a variety of methods based on the nature and level of the available data. The most common method for correlation with interval data is "Pearson's *r*," named after Karl Pearson. This remarkably flexible statistical tool can detect relationships in different kinds of variables, like the length of the big toe and intelligence, or the number of bumps on the head and the nature of personality. You may recognize these examples as real ones in history! Phrenology was a very serious study at one time, but recent advances in psychology and physiology have eliminated cranial bumps as a serious indicator of human character.

## STRENGTH AND DIRECTION

Pearson's *r* yields values that vary from (−1.00) to (+1.00). The *direction* (negative or positive) of the correlation is indicated by the sign (− or

+), and does not mean bad or good. The *strength* of the correlation is indicated by how close to +1 or −1 the number is.

Thus, for example, if we have a −.78 relationship between the length of a work shift and the production rate, this would indicate that:

1. The direction is negative—inverse—so that as the shift length increases, the production rate decreases.
2. The relationship is fairly strong because it approaches (−1.00).

The top panel of Figure 4.4 shows a perfect negative correlation, since each increase in the shift length is exactly matched by a one unit decrease in the production rate. As this occurs, all the scores fall on a straight line with a downward slope. In reality, we never observe a perfect negative or a perfect positive correlation.

Correlations approach or move further away from zero. The bottom panel of Figure 4.4 is more realistic because people differ in their productive ability and stamina. There is still the downward (negative) slope, but the scores spread out a bit, and the correlation of (−.78)

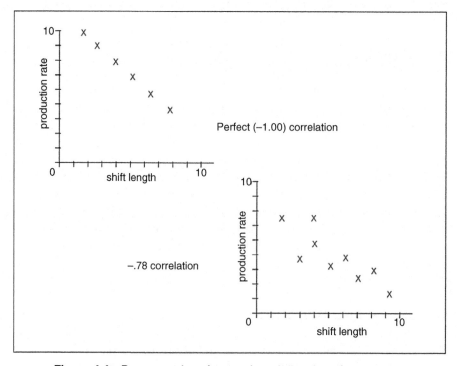

**Figure 4.4**    Demonstration of strength and direction of correlation.

approaches (–1.00). To repeat, *negative* is simply the direction, whereas the *strength* of the relationship is determined by how close $r$ is to (–/+ 1.00). We will discuss a bit later about how to judge the strength of a correlation.

Figure 4.5 shows an actual example of a correlation scatter diagram from SPSS® output with a correlation of (–.62), a moderately strong negative correlation. The variables shown are from total or "aggregated" school scores in the state of Washington. The outcome variable on the vertical $y$ axis represents the percentage of fourth grade students by school who met the passing standard on the state level math test, and the predictor variable on the $x$ axis is the percentage of students eligible for "FreeorReducedPriceMeals" (F/R), a proxy for family income level. This figure shows how family income and fourth grade math achievement are related at the school level for over 1000 schools. The scatter diagram is hazier than the sample graphs in the former figures because there are so many more data points, but a general trend is present in the scores, each of which represents a single school. You can detect a general downward scatter of scores that are heavier toward the middle of the group with several outlying scores.

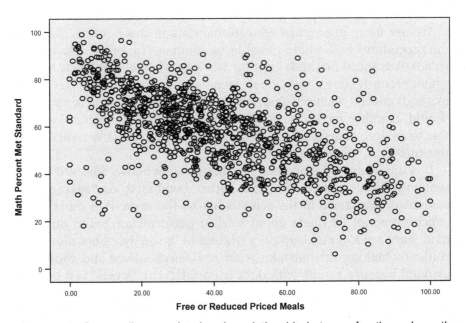

**Figure 4.5** Scatter diagram showing the relationship between fourth grade math achievement and family income by school.

The scatter diagram shows that higher math achievement scores are related to lower F/R rates and lower math achievement scores are related to higher F/R rates. This means that schools with higher percentages of children with lower family incomes show lower fourth grade math achievement scores. This is a bit difficult to interpret since we are using F/R percentages that are *higher* when more children qualify for free/reduced lunch due to *lower* family incomes. (In this instance high F/R is an indication that more children at a school have lower family incomes.) Overall, the data suggest a moderately strong inverse relationship between family incomes and math achievement scores.

## A NOTE ON THE NATURE OF THE DATA

We will have more to say about our data in later sections as we work with specific features of the data set. The entire database was downloaded from the website of the department of education in the state of Washington and can be explored as a learning tool to understand school-level achievement in the state.[3] The state database was vetted by their data managers for issues like nonreporting for categories consisting of less than 10 students. You can find similar data available for other states by exploring their department of education websites.

We are using *aggregated* educational data in this book to illustrate the procedures and, where possible, to comment on broad educational trends. Advanced methods exist for working with several levels of data simultaneously (e.g., individual level and school level), but these methods exceed what we describe in this book. Interested researchers should examine *hierarchical modeling and multilevel regression* as a way to analyze data simultaneously from different levels. We will discuss this further in a later section.

It is important to remember that the data set above does not represent individual student data, but rather combined, or "aggregated," student performance in the schools across the state. It is crucial that conclusions not be made about *student* performance based on these data, since *schools* make up the variables in the analyses, not individual students. Making this mistake is an *ecological fallacy* and should be avoided because results with data from different "levels" can be very misleading.

[3] The data were downloaded from the website of the Office of the Superintendent of Public Instruction in Washington (http://www.k12.wa.us/) with permission.

I might also mention that the correlation will look different depending on the data used. That is a very important consideration for all of statistics. In this example we used "aggregated" scores or individual scores that were summed up across the entire school. Thus we observed the percentage of students eligible for subsidized meals in the school and the percentage of students who met the math standard at the school. This resulted in two "continuous" scores for each school; scores that can take a range of values on a scale where we assume that the differences between the values are equivalent across the scale (weight, income, etc.). We then compared all schools in the database on these two variables using Pearson's correlation.

The analyses that follow using these data are based on fourth grade data in schools. In Washington state, the state achievement test is administered at several grade levels. While results involving family income data are very similar for fourth and other grades, there are some indications that comparable analyses for higher (e.g., tenth grade) do not have the same validity. Therefore, because we are using the data as a way to understand the statistical methods, we will focus only on fourth grade results.

A common reason for using aggregate data in educational studies is that little individual student information is available for educational researchers due to concerns over privacy. In addition there is typically no common information gathered by school districts about family income other than whether or not the student is eligible for subsidized meals. Thus each student in our database was classified by school districts as either eligible or not eligible for subsidized meals.

Using student-level information for this analysis, if it were available, would result in correlating students who were either eligible or not with students who either met the standard or did not meet the standard. Strictly speaking, this would require a different statistical process, as the "level" of the data would be different from the continuous percentages that we used. Since we would have "either/or" scores on both variables, this would require that we use a correlation method designed for "categorical" or "nominal" data where the information is presented as categories that do not necessarily have a numerical relationship or that do not have equal distances within each scale. An example of a categorical scale would be one's chosen academic major in college as chemistry, sociology, and so forth. A "dichotomy" is a situation where categorical data have two categories. We would have dichotomous scores if we were to have used individual student data described above.

Another issue that might have affected the outcome of our study was the data used for math achievement. In our example we used the percentage of students at the school that passed the state standard for mathematics. This means that we used "criterion-referenced" scores, or scores measured against an established standard of performance. Alternatively, we could use "norm-referenced" scores, where individual student scores are compared to one another. If we had used math "scale scores," we would have observed an overall correlation of −.473, markedly lower than the −.622 we observed with the other measure. There are thus differences between these two sets of measures. We chose the measures we did because we were concerned about the extent to which individual differences in each school would be masked if all the individual scale scores were combined in a school average.

## INTERPRETING PEARSON'S *R*

The actual correlation we examined above (−.62) appeared to be moderately strong. An important question for interpreting the results is, How strong is strong? We can determine that the correlation is statistically significant, but how do we know if it is meaningful? Stated differently, the correlation is moderately strong compared to what? The correlation of −.622 is closer to 1 than it is to 0 so that is an indication that the relationship between the variables tested is likely meaningful. But are there other ways of deciding how strong, or meaningful the resultant *r* values are?

Several issues are very important to evaluators who use statistics to make decisions. As we noted earlier, "statistical significance" refers to the extent to which we can assume the results we obtained are not likely to have come from a population in which the "real" or actual correlation is 0. That is, we want to be confident that our finding from the scores we have available is not a "fluke" finding, or a result of chance. Thus the first important consideration over the strength or meaningfulness of the results is to determine whether or not we can have confidence that they are not an "accident" or occur by chance.[4] We do that through a process of testing the hypothesis that our observed correlation is more than a chance finding.

---

[4]In the case of math achievement and F/R, using this definition of significance may be awkward because we are actually testing the correlation of the variables for ALL schools in the state. Therefore assessing the likelihood of our "sample" correlation being from a population where the "actual" population correlation is 0 is strained, since we are calculating the actual population correlation. In this instance we are therefore viewing descriptive statistical findings, not inferential processes.

In assessing the meaningfulness of a correlation, we will rely on the response to two questions. First, is the correlation *statistically significant*, as we discussed above. Second, is the correlation *practically significant?*

## TESTING THE STATISTICAL SIGNIFICANCE OF A CORRELATION

As we noted in an earlier section, concluding that a correlation is statistically significant involves assessing the likelihood that a sample (or our obtained) value is likely to come from a population where there is no relationship between the variables. This assessment relies on an hypothesis test where we state the null hypothesis and then reject (or accept) this statement, depending on a set of tabled comparison or "critical rejection" values for correlation. This is one of the earliest lessons in inferential statistics. If the calculated correlation is large enough and exceeds the critical value for rejection, it will likely be found not to be a result of chance, and we can reject the null hypothesis.

In procedural terms we are "stacking up" our sample findings with what we assume the actual population value might be. Is our sample population likely to be found in the population from which it was taken, or is it so different or changed that it can no longer be considered part of the original population? If the null hypothesis assumes that there is no correlation in the population, and our sample values find there is a non-0 correlation, does that mean the sample correlation is a result of chance, or fluke, or some oddity, or does it mean that our sample values are so different from an assumed 0 population value that the sample can no longer belong to that population? If it is the latter, then we can assume our sample correlation reflects a "real" relationship between the variables we tested.

The only caveat to these findings would be a consideration of "alpha error," which indicates the probability that rejecting a null hypothesis can be done in error. That is, a statistically significant outcome may in fact represent one of the small probabilities that it is the result of chance. Statisticians usually settle on this small probability as .05, or 5 times out of 100 that a finding may be the result of chance rather than the observed relationship of a calculated correlation.

As with other statistical tests, we need to make reference to a table of values to help us establish whether the observed, sample correlation is so large (given the number of cases) that we can reject the assumption that it came from a population with a 0 correlation.[5] If our obtained correlation exceeds the tabled value, we can conclude that the sample $r$ is not a product of chance and represents a real relationship. Using the table of values (which are available in most any statistics textbook), we would determine whether our sample correlation exceeded the critical value (either as a positive or negative value) and could then be said to be "statistically significant." As an example, our (−.62) correlation above would be considered statistically significant, since the tabled value for an analysis of over 1000 cases would only need to exceed .08 correlation to be considered significant (at the $p < .01$ level of significance).

Since most evaluators no longer need to compute correlations by hand, they have less need of referring to tables of values that identify the critical values of rejection. We use statistical software programs that do this evaluation for us. In the SPSS® case the program calculates the *actual* significance level rather than identifying what "level" of significance was obtained or surpassed.

You can use SPSS® to perform correlation analyses by making selections from the drop-down menus. Using the example discussed above, we can create a correlation analysis by choosing "Analyze–Correlate–Bivariate" from the SPSS® menu system as shown in Figure 4.6. Selecting "Bivariate" will yield another menu in which you select the variables from the database you wish to use in your analysis, as shown in Figure 4.7.

Table 4.1 shows the SPSS® output table for the correlation analysis that would be generated from this procedure. Earlier we observed a scatter diagram that showed the pattern of the data between fourth grade math achievement and F/R rates in Washington schools (Figure 4.5). The SPSS® output in Table 4.1 relating to this analysis is a "matrix," since it presents all variables correlated to each other. Thus both variables are located on the columns and the rows. For example, the correlation between "MathPercentMetStandard" and itself is 1.0, a perfect correlation. This is to be expected since every value of math achievement for every school is exactly the same. The same observation can be made when correlating F/R with itself. Both of these figures lie on the "diagonal," extending from the top row and column to the bottom row and column. Next math achievement is correlated with F/R in two places. Both values are equal, since it doesn't matter how you

---

[5] In practice, we rarely use reference tables for making these decisions, since the information is provided by statistical analysis programs. The SPSS® program directly calculates and reports the actual probability obtained of the result.

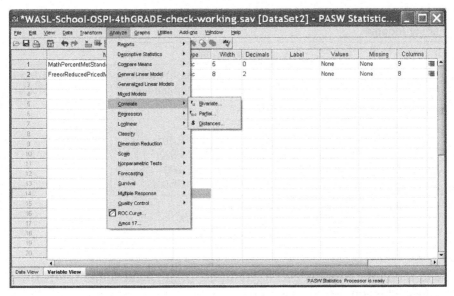

**Figure 4.6**    SPSS® screen showing the menu to identify a bivariate correlation.

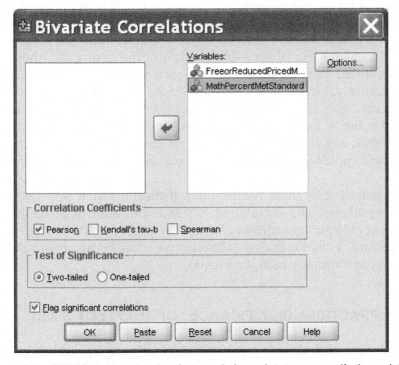

**Figure 4.7**    SPSS® screen showing the correlation submenu to specify the variables to be used in the analysis.

**Table 4.1  SPSS® Correlation Output for School-Based Achievement and Family Income**

Correlations

|  |  | MathPercent MetStandard | FreeorReduced PricedMeals |
|---|---|---|---|
| MathPercentMet Standard | Pearson Correlation | 1 | −.622(**) |
|  | Sig. (2-tailed) |  | .000 |
|  | N | 1050 | 1050 |
| FreeorReduced PricedMeals | Pearson Correlation | −.622(**) | 1 |
|  | Sig. (2-tailed) | .000 |  |
|  | N | 1050 | 1050 |

**Correlation is significant at the .01 level (2-tailed).

read the table: row 2 (F/M) to column 1 (math achievement), or row 1 (math achievement) to column 2 (F/M).

It is also clear from the output that the SPSS® program indicates that the correlation between F/M and math achievement has surpassed the .01 level of significance, indicated by the two asterisks and the related footnote. This is a finding similar to our discussion above of a hypothesis test, which concluded that a sample value surpassing the tabled value for the .01 level of significance. The matrix also shows the rows entitled "Sig. (2-tailed)" under each variable. This is the *actual* significance value calculated by the SPSS® program. Since this value is reported only to three decimal places (.000), we don't know what the actual value is, but we can be certain that it is "beyond" (or smaller than) .001, which indicates that there is less than one chance in 1000 that our sample value would come from a population with a real $r$ value of 0.

As indicated by these results, the statistical decision is to reject the null hypothesis—the correlation is significant. Statistical significance does not indicate how *meaningful a result might be*; just that the result is likely not a chance finding (although we could be wrong in rejecting the null and commit an alpha error!).

## THE "PRACTICAL SIGNIFICANCE" OF *R*: EFFECT SIZES

Recall an earlier discussion that indicated more ways to determine "how strong is strong" than referencing the statistical significance alone. If we conclude that a sample correlation is statistically significant, we can use another method to determine how meaningful the correlation is for the research problem. Researchers refer to this as determining the "effect size" of the resulting correlation.

Simply looking at the size of the correlation gives us no real sense of how strong the relationship is between two variables, so we have to approach the question of meaningfulness or effect size another way. One way of making this assessment in correlation procedures is to square the obtained correlation. In the example we just used, the square of the −.62 correlation is .38. Statisticians refer to this value, the square of the correlation coefficient, as the "coefficient of determination." The square of *r* refers to the amount of variance in one variable accounted for by the other.

What this means is this: we can consider the fact that a distribution of scores (e.g., math achievement test scores) varies a certain amount, or spreads out around a mean score. The question is *why* do the scores spread out or vary? In a world where every child was the same, there would be no variance—each child would get the same score. Something, or things, is responsible for children getting different scores. The correlation between variables provides a partial explanation for this variation. By squaring the correlation, we can understand this percentage of "variance explained" in the outcome (i.e., math achievement) as being accounted for by a predictor variable (i.e., eligibility for free or reduced price lunch).

The (−.62) correlation between schools' math achievement and proportion eligible for F/R therefore has a coefficient of determination of approximately 38%. This indicates that 38% of the variance in school math achievement is accounted for by the proportion of students eligible for F/R. The next question concerns the magnitude of the effect sizes—how large do they have to be to be considered meaningful?

Another example from my evaluation projects will help to illustrate and clarify the magnitude of effect sizes. In this study we were concerned about possible connections between achievement, study time, reading time, time spent watching TV, and so forth. With a large database of ninth grade student achievement scores, we correlated two variables on another common achievement test, the ITBS (a state achievement test): the total reading achievement scores and the time per week the student spends reading. The resulting correlation was .27, and was statistically significant. The question was, So what? How can we determine whether a correlation of .27 is strong or weak? By squaring the correlation, we can see in Figure 4.8 the measure or amount of the variance in one variable is accounted for by the other. This component of the variation" will be helpful in the interpretation

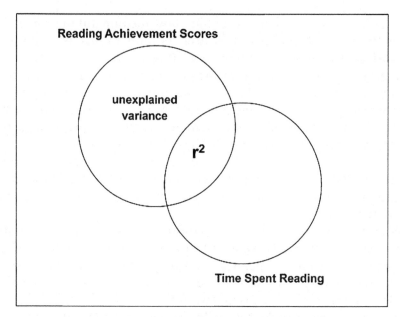

**Figure 4.8**   Accounting for variance through correlation.

of the analysis of the prediction of the value of one variable by using the other variable, where both are significantly related.

If we establish a correlation (e.g., between achievement test scores and time spent reading), the variables' measures overlap, as shown in the Figure 4.8. We can understand this overlap conceptually to be the amount of variance in the original spread of the scores (reading achievement) that is accounted for by this new variable (time spent reading). We can conclude, for example, that knowing how much reading a child does is a partial explanation of why he or she achieves a specific score. This analysis doesn't explain everything, since there is still a lot of "unexplained variance," but it "chips away" at the overall spread of the scores. The $r^2$ figure is this "accounted for" variance shown as the overlap in Figure 4.8.

To go back to our example, if we square the original correlation of .27, we get a figure of approximately .07. This means that 7% of the variance in reading achievement is accounted for by how much time ninth grade students spend reading per week. We have already determined that the .27 correlation is statistically significant, but how much of reading achievement can be accounted for by time spent reading? In this case the answer is 7%. Is that a lot or a little?

Some evaluators suggest that determining the meaningfulness or impact of one variable on another is a matter of understanding the context of the research. Thus, with only one variable correlating with another, 7% might be helpful as we continue to gather a broader understanding of reading achievement. However, the absolute amount of 7% doesn't appear to be much. If we were researching a topic about which very little is known, like some exotic disease, however, we might be ecstatic to find one variable that explains this much!

Recognizing the uncertainty of this situation, some researchers have proposed "cutoff" values for determining whether or not the $r^2$ amount is meaningful. These cutoff values vary, but one example from Cohen et al. (2003, p. 93), is a range of .02, .13, and .26 for small, medium, and large effects, respectively. Thus, in our example, 7% would represent a small to medium effect. In a research situation, this would probably not be evidence to announce to the press, but it might be a helpful finding that is part of a larger investigation of the determinants of reading achievement. In such a wider investigation, we would undoubtedly use additional variables to help us explain reading achievement rather than reading time alone. We might consider such variables as student aptitude or interest along with considerations of parental involvement, nature of the reading curriculum, socioeconomic status, and others. In any case, the determination of how meaningful a certain variable is for explaining another relies on more than a statistically significant finding. Evaluators must use their judgment as well.

## AN EVALUATION EXAMPLE OF CORRELATION: THE IMPACT OF TECHNOLOGY ON TEACHING AND LEARNING

In many of the studies we performed on TAGLIT data, we created indexes from the responses and then examined the correlations among these indexes from teachers and students. We used factor analysis (FA) techniques to create these indexes with SPSS®. FA (and the related principal components analysis) techniques allow a researcher to discover relationships among "clusters" of test items on an instrument that are highly related to one another, but have lower relationships to other clusters.[6] The assumption is that each test item reflects a certain analytical meaning on the part of the respondent that can be detected by its relationship to other items. Thus the evaluator can "inductively"

---

[6]FA and PA are straightforward on a conceptual level, but can be difficult to conduct and interpret. The reader who wishes to pursue these techniques might seek references that are devoted entirely to explaining the process. A good sketch of the processes is provided by Grimm and Yarnold (1998).

identify groups of items within an overall survey instrument with similar underlying meaning and then assign a name to the group of items that reflects this meaning. By using these techniques, the evaluator can observe which items "cohere" into a single meaningful "factor" and then create an index (either by summing or averaging the item responses) that represents all the constituent items. Among the practical uses of this process is that you might use a single item from the clusters of related items to "represent" the group of items as a way to create a shorter instrument for subsequent testing.

One of our studies that used FA and correlation addressed whether or not there was a relationship between the extent to which middle/high school teachers used classroom technology ("skills") and their approach to teaching and learning ("impact") in the classroom. We used PA to create the "skills index" by assessing "how far along" the teacher was in terms of learning to use various technologies (word processers, spreadsheets, presentation software, and the like). The "impact index" reflected whether, as a result of their use of technology in teaching and learning, teachers would involve students in different aspects of the classroom (e.g., cooperative rather than competitive learning, activities that required higher level thinking skills, interaction with the outside world, interdisciplinary activities) and work with students who needed extra help, "coach" rather than lecture, and the like.

Using SPSS®, we determined that there was a significant correlation between the two indexes among both elementary and secondary teachers when they were aggregated to the school level. This 2003 database consisted of over 121,000 middle and high school teachers' responses and almost 137,000 elementary teachers' responses. When aggregated, this represented over 4400 and 6800 schools (respectively) nationwide and indicated several important facets of the relationship between teachers' technology use and the nature of classroom learning.

To see how the middle/high school teacher analysis used SPSS®, look at the following figures, which show how the evaluator can use the Analyze menus to conduct the correlation analysis. Figure 4.9 shows the menus in which you select Correlate from the Analyze drop-down menu, and you will see choices for the correlation analysis. Choose Bivariate correlation, which is simply a two variable correlation. As you can see in Figure 4.10, you can specify Pearsons, Kendall's tau-b, or Spearman's correlation. Since the TAGLIT data are interval level data, you can use Pearson's correlation. It is the default choice by SPSS®. You will also note in Figure 4.10 that the default by SPSS® is to "Flag significant correlations" and to select "Two-tailed" tests of significance. Figure 4.10 also shows that I selected two indexes for the

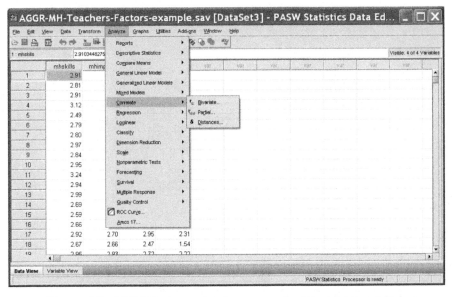

**Figure 4.9**  SPSS® screen showing menus for bivariate correlation.

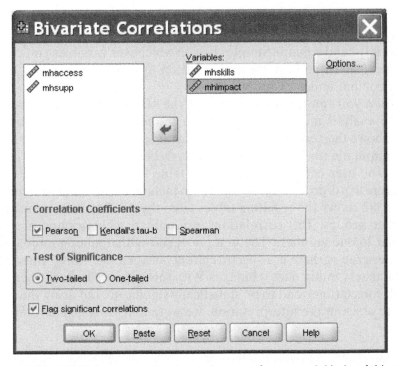

**Figure 4.10**  SPSS® screen showing submenus for several kinds of bivariate correlations.

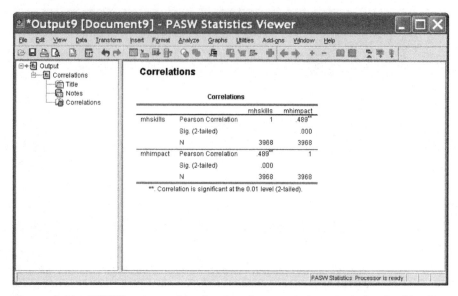

**Figure 4.11** SPSS® output table showing the correlation matrix for the bivariate correlation.

analysis: "mhskills" and "mhimpact" from the list of available variables in the left-hand column. Simply click on whichever variables you choose for the analysis and use the arrow button to move the variables to the right column under "Variables."

When you run the analysis (select the "OK" button on the bottom of the analysis menu screen), SPSS® creates a separate "output" file that shows the results of the analysis you specified. Figure 4.11 shows the output file for the correlation analysis between mhskills and mhimpact. The user can select the output table and import it directly to a separate word-processing document to include in the evaluation report. Table 4.2 shows the resulting tables from the SPSS® analyses for both teacher groups. The correlations for both teacher groups are very similar to one another.[7] The first panel shows that among middle/high teacher groups, there is a .49 correlation between the skills ("mhskills") and impact ("mhimpact") indexes. With this large a database, even very small correlations tend to be statistically significant, but using the effect size to assist in the interpretation, we were able to conclude that 24%

---

[7] It is important to remember that these are school-level results rather than individual-level results, so the data should be interpreted as "index correlations among teachers in m/h schools..." for example.

**Table 4.2   TAGLIT Correlation Results for Middle/High and Elementary School Teacher Groups with Skill and Impact Indexes**

| Correlations | | | |
|---|---|---|---|
| | | mhskills | mhimpact |
| mhskills | Pearson Correlation | 1 | .489[**] |
| | Sig. (2-tailed) | | .000 |
| | N | 3968 | 3968 |
| mhimpact | Pearson Correlation | .489[**] | 1 |
| | Sig. (2-tailed) | .000 | |
| | N | 3968 | 3968 |

**Correlation is significant at the .01 level (2-tailed).

| Correlations | | | |
|---|---|---|---|
| | | eskills | eimpact |
| eskills | Pearson Correlation | 1 | .487[**] |
| | Sig. (2-tailed) | | .000 |
| | N | 6142 | 6142 |
| eimpact | Pearson Correlation | .487[**] | 1 |
| | Sig. (2-tailed) | .000 | |
| | N | 6142 | 6142 |

**Correlation is significant at the .01 level (2-tailed).

$(.49^2)$ of the variation in classroom teaching and learning could be accounted for by the middle/high school teachers' learning to use the various technologies.

The second panel of Table 4.2 shows a .49 correlation between the indexes ("eskills" and "eimpact") among elementary teachers' schools. This translates to a 24% effect size, or the skills index accounting for almost one-fourth of the variance in the impact index.

Both results show that technology skills of both teacher groups are helpful for explaining the use of technology in the classroom. We should note that the explanation could be reversed as well. That is, that variation in technology skill is accounted for by technology use in the classroom, but the meaning of the variables would seem to favor the previous interpretation.

This result, while simple at one level, can have important meaning for classroom learning. To the extent the correlations above are representative of other teacher groups in other schools, one could suggest that improving teacher skill with various technological applications might be helpful in ensuring that they are used in student learning approaches that favor cooperative learning rather than competitive learning, activities that require higher level thinking skills, interaction with the outside world, interdisciplinary activities, and the like.

While we cannot conclude a causal relationship between these variables, we can nevertheless point to a potentially meaningful discovery for better understanding the nature of teaching and learning. We used correlation in this study to discover an important finding embedded in the survey data. Certainly there are other factors that could add to the understanding of what impacted classroom learning, but this method illuminated one very important determinant. In our study we identified two further indexes that provided additional insight to our analyses. The TAGLIT instruments included items that asked teachers about the extent to which technology (word processing, database/spreadsheets, email, WWW, etc.) was available to teachers that resulted in the "access" index. The instruments also identified teacher perceptions of the availability of support personnel for setting up, repair, and maintaining hardware, and teaching staff members how to use technology, resulting in the "support" index. Table 4.3 shows the results for the elementary school teacher's schools.

The results for the table are in the form of a matrix reported by SPSS®. Each of the variables (indexes) is correlated with each of the others, and the results include the Pearson correlation coefficients along with their significance levels and the number of cases on which the analyses are based. Table 4.3 shows significant correlations among all the correlation pairs. The "eskills–eimpact" correlation is .49, which we discussed above. The technology skills index is also significantly related to technology access ("eaccess"), where the correlation is .40. Thus about 16% ($.40^2$) of the variance in skills is due to access of technology, and vice versa. Table 4.3 also shows that technology skills is significantly correlated to technology support, with $r = .26$, and an effect size of .07 ($.26^2$). All the correlations among the indexes are in fact significant. The subsequent panel shows similar results for the MH teacher groups.

The results in Table 4.3 indicate a number of important findings, but they also point out issues that evaluators face in making conclusions about the analyses they perform.

*Correlation and Effect Sizes*  When outcomes are based on sizable samples, even small correlations may be statistically significant. In these cases, the evaluator needs to consider the effect sizes as a way to assist in reflecting on the nature of the findings. In the results shown in Table 4.3, for example, the evaluator may recognize, using the criteria we discussed above for evaluating effect sizes, that the

**Table 4.3   Correlation Matrix for Elementary and Middle/High School Teacher TAGLIT Factors**

| | Correlations | | | |
|---|---|---|---|---|
| | | eskills | eimpact | eaccess | esupp |
| eskills | Pearson Correlation | 1 | .487** | .404** | .255** |
| | Sig. (2-tailed) | | .000 | .000 | .000 |
| | N | 6142 | 6142 | 6142 | 6142 |
| eimpact | Pearson Correlation | .487** | 1 | .350** | .296** |
| | Sig. (2-tailed) | .000 | | .000 | .000 |
| | N | 6142 | 6142 | 6142 | 6142 |
| eaccess | Pearson Correlation | .404** | .350** | 1 | .642** |
| | Sig. (2-tailed) | .000 | .000 | | .000 |
| | N | 6142 | 6142 | 6142 | 6142 |
| esupp | Pearson Correlation | .255** | .296** | .642** | 1 |
| | Sig. (2-tailed) | .000 | .000 | .000 | |
| | N | 6142 | 6142 | 6142 | 6142 |

**Correlation is significant at the .01 level (2-tailed).

| | Correlations | | | |
|---|---|---|---|---|
| | | mhskills | mhimpact | mhaccess | mhsupp |
| mhskills | Pearson Correlation | 1 | .489** | .357** | .186** |
| | Sig. (2-tailed) | | .000 | .000 | .000 |
| | N | 3968 | 3968 | 3968 | 3968 |
| mhimpact | Pearson Correlation | .489** | 1 | .318** | .241** |
| | Sig. (2-tailed) | .000 | | .000 | .000 |
| | N | 3968 | 3968 | 3968 | 3968 |
| mhaccess | Pearson Correlation | .357** | .318** | 1 | .634** |
| | Sig. (2-tailed) | .000 | .000 | | .000 |
| | N | 3968 | 3968 | 3968 | 3968 |
| mhsupp | Pearson Correlation | .186** | .241** | .634** | 1 |
| | Sig. (2-tailed) | .000 | .000 | .000 | |
| | N | 3968 | 3968 | 3968 | 3968 |

**Correlation is significant at the .01 level (2-tailed).

effect sizes for the correlations between skills and impact is large, the correlation between skills and access is medium, and the correlation between skills and support is small. This is despite the fact that each of these correlations is statistically significant.

*Causal "Direction"*   Assessing the direction of influence in the relationship from correlation results is often very difficult. As we saw above, skills and impact could be considered mutually determinative, although it makes some initial sense to say that having the skills influences their use. What of the relationship between skills and access, however? It would not necessarily follow to say that skills influences access. It may appear more likely that having access to technology might have a bearing on how much skill teachers have with using technology, although that may not be plausible either. Just because technology is available does not mean that people will develop the skills to use it. The same arguments could be made when examining the relationships between skills and technology support offered by the school. The effect sizes may be one clue to the nature of the relationships and how to interpret the findings. The skills–impact relationship has a strong effect size, which may lend itself to a bit more clarity on the direction of the relationship.

*Limitations of Correlation Findings* It is important to analyze the other relationships in the matrix that may help us reach some tentative conclusions about the directions of the relationships. Table 4.3 shows a very high correlation between the "support" and "access" indexes among the elementary teacher groups, with $r = .64$. This yields an effect size of 41% ($.64^2$), considered high by the criteria we discussed earlier. In this case it may appear easier to understand the direction of influence; both indexes are mutually determinative. Teachers' perceptions of having access to technology elements in the schools are related to their perception of support for technology. The impact–access relationship, with $r = .35$ has an effect size of about 12% ($.35^2$), considered small to medium. Here it might appear logical to interpret the results as having access to technology might result in greater impact on teaching and learning (impact).

All these relationships point out one of the primary deficiencies of correlation analysis for the evaluator. Correlation findings are very powerful and helpful for illuminating relationships, but there is a limit to how the relationships can be interpreted. In a correlation pair, either variable may be determinative, so the evaluator needs additional tools to be able to discover the direction of influence. The influence of a third, or more, variable(s) may not be apparent when only examining a two-variable relationship. There are several statistical tools available to assist with this dilemma. Foremost among these is the use of regression, which this book develops in great depth. Using regression and other statistical techniques can move the evaluator from examining correlation pairs to examining all the variables together to see what the directions of influence are, and the extent to which each is determinative.

We used correlation analyses for additional insights in this study by relating student results to teacher results, and adding several of the school leader results to the analyses. Our study used these as the primary insights to launch a more detailed study of all the variables taken together using multiple regression. We will revisit these findings after we discuss the nature and use of multiple regression.

## INFLUENCES ON CORRELATION

The world of evaluation is one fraught with limitations to data, oddities, and other considerations that we must take into account when performing statistical procedures. It is important to look carefully at these before you carry out the correlation analysis. Depending on the peculiarity of the data, the resulting correlation could be dramatically different.

### Restricted Range

One problem to recognize is "restricted range," which can impact the strength and size of a correlation. In this case an evaluator might correlate only a segment of a sample because that might be all that is available at the time, or for some other reason. To take an example, suppose that an evaluator wanted to examine the relationship between math achievement and SES, as we did with an earlier example. However, this time, because of the sample of schools we were studying, the evaluator only collected data on students from schools where the F/R value was high, that is, schools with high numbers of students qualifying for subsidized meals.

Figure 4.12 shows the resulting scatter diagram where we use data from schools in the highest fourth of the F/R percentage in the data set, those where the F/R is 58.83% or higher. The calculated correlation in this instance is −.318, quite a bit lower than the −.622 correlation shown earlier. The resulting coefficient of determination is .10, compared to the .39 of the overall database. By selecting only a segment of the overall sample of schools, the evaluator would end up with two very different results. Compare the scatter diagram from Figure 4.13 that contains all the data. You will see that the "restricted range" scatter diagram represents a relationship that is still statistically significant, but the pattern is much less discernable than the scatter diagram that contains all the data.

If an evaluator has access to a sample that is relatively homogeneous on the variables (all the subjects appear to be similar in achievement,

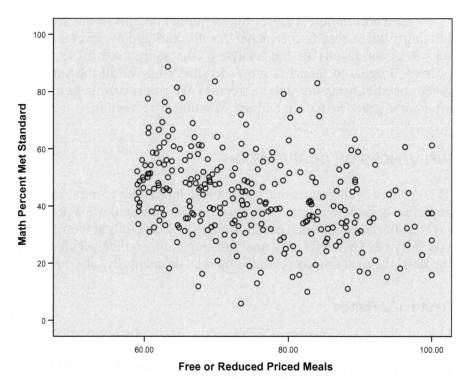

**Figure 4.12** Scatter diagram showing the effects of restricted range.

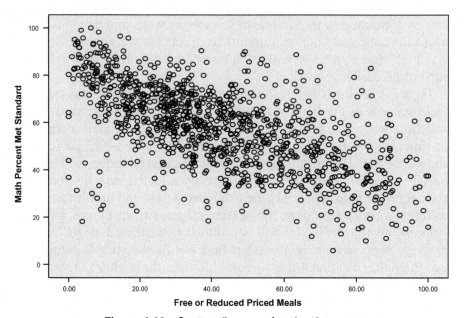

**Figure 4.13** Scatter diagram of entire data set.

SES, etc.), the resulting correlation can appear to be lower than it is with a wider range of scores. In Figure 4.13, if students from schools of all different achievement scores and time reading amounts were correlated, there may be a strong correlation. But if we only have students from schools who are very similar to one another, like a group of students from schools with gifted reading programs that stress obsessive reading, the resulting correlation would be less pronounced, and perhaps not even be significant. Evaluators must remember to ensure that they have a full range of scores in the research sample.

## Extreme (Outlier) Scores

Another problem in actual research is to determine whether or not there are "outlier" scores that might affect the correlation. Outliers are scores that deviate widely from the overall pattern of scores and may affect the resulting calculation of the correlation. You can easily imagine how this could happen. In our example above, a school might submit achievement scores or F/R scores that were compiled incorrectly. Or, the data entry worker could have inadvertently entered an incorrect value into the spreadsheet. Fortunately, there are several ways to detect scores that vary widely from the others and to make an assessment about whether they are "actual" scores or mistakes.

Computer software programs like SPSS® have several methods for isolating extreme scores. Once detected, the evaluator must determine what to do with these scores. Some scores that appear to be outliers may represent actual scores and should therefore be included in the analysis. Others might be eliminated from the analysis, replaced with the average score in its distribution, or replaced by another value according to an a priori rationale. We will have more to say about detecting extreme scores in a later chapter.

## OTHER KINDS OF CORRELATION

There are several different correlation calculations, depending on the nature of the data available to an evaluator. We have discussed data that we assumed were "interval" data where the distances between the values were equivalent across the range of our data. (Thus, for example, the difference between 10% and 11% F/R is the same as the distance between 32% and 33% F/R.) This is an important matter because, if the distances are not equal, then we cannot meaningfully use addition, subtraction, multiplication, and division as these procedures require that all the values be based on equivalent, known distances.

If data do not meet the equivalent distance criterion, evaluators can still use correlation analyses to help discover patterns embedded in the data. One of the most prevalent of these is "rank order" correlation, where data include only values that can be considered greater/less than each other, but without necessarily being equally distant on a number scale. For example, we might want to examine the relationship between teacher appraisal of students and student success in the classroom. In this instance we might have the teachers' ranking of students from "highest performing" to "lowest performing" using designations of "first, second, third … , etc.," and the class standing based on homework assignments. Notice that both variables are rank ordered, but without the distance between the ranks assumed to be equal among students. In such a case the evaluator might use *Spearman's correlation*, which is based on ranks. Using our earlier example, we can examine the correlation between school-level math achievement and F/R percentage. When we conduct both analyses using SPSS®, both the Pearson's and Spearman's correlations are (−.62).

There are many different ways to calculate correlation—depending on the nature of the research problem and available data, whether they are interval or ordinal, categorical or continuous, and so on. Some of the alternatives include Kendall's tau, biserial correlation, the phi coefficient, and the contingency coefficient. A very important point to emphasize is that evaluators *must use the appropriate tool for the nature of the situation presented for the study.* Using the wrong measure for the available data will result in less powerful results and possibly misleading conclusions. Before running a correlation analysis, be sure you have investigated the nature of the data and are using an appropriate procedure. Table 4.4 presents a brief view of some of the correlation procedures available from SPSS®:

There are many other correlation techniques, each tailored to a research situation in which the data are of a specific type. Be sure to

**Table 4.4   Examples of Correlation Techniques Available in SPSS®**

|                          | Variable $X$ | Variable $Y$ |
|--------------------------|-------------|-------------|
| Pearson's correlation    | Interval    | Interval    |
| Spearman's correlation   | Ranks       | Ranks       |
| Kendall's tau-b          | Ranks       | Ranks       |
| Contingency coefficient  | Categorical | Categorical |

consult appropriate resource material for the procedure you plan to use so that you can properly interpret the results.

## A Research Example of Spearman's rho Correlation

The following is an example of Spearman's correlation, which is correlation that is based on ordinal, or ranked, data. Such analyses are termed "nonparametric" because they do not refer to population parameters. Pearson's correlation is calculated with reference to the attributes of the (assumed) population distribution. With less than interval data, researchers typically conduct nonparametric statistical tests because they do not make these assumptions. They are derived directly from the data at hand.

With TAGLIT data we can use Spearman's correlation to help us understand the relationship between the "urbanness" of a school and the student–teacher ratio. Are rural schools more likely to have larger or smaller classes? School leaders provide data on schools using the TAGLIT instruments, including a classification of how urban or rural the area of the school, and school features such as the student–teacher ratio. The "metro3" variable classifies schools as "urban," "sub-urban," or "rural" according to U.S. Census Bureau designations of urban area populations. Since the categories are used for broad classification, we cannot make the assumption that they represent interval data. However, we can assume that the categories represent rank order data, since the "rural" designation would be less populated than the "suburban" or "urban" designations. The metro3 variable assigned values of "1," "2," and "3" to urban, suburban, and rural, respectively.

Figure 4.14 shows how we would use the SPSS® menu to create a correlation analysis using Spearman's correlation. The SPSS® procedures in Figures 4.6 and 4.7 show the "Analyze" menu wherein you would select the "correlation" procedure, and then the "bivariate" procedure, since we are dealing with two variables. As you can see from the Figure 4.14, the bivariate correlation panel would allow you to specify the kind of correlation you wish to use. I selected the boxes for "Pearson" (used for interval data), "Kendall's tau-b" (typically used when both variables are ranked), and "Spearman." Using the arrow indicator, you can move the variables of interest to the "Variables" window that will show you which variables will be analyzed. Choosing the "OK" button at the bottom of the panel will produce the SPSS® output file showing three different sets of correlation results.

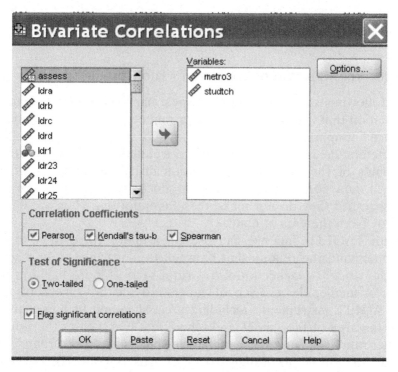

**Figure 4.14** Bivariate correlation submenu for TAGLIT analyses.

Table 4.5 shows the results for the Kendall's tau-b and Spearman's rho analyses. In both cases the correlation between metro3 and studtch (student–teacher ratio) are significant. This is not a surprise, since we have data on such a large number of schools in the study ($N = 6615$ and 10021 for metro3 and studtch, respectively). In both cases the correlations are inverse, indicating that the more urban the school, the greater is the student–teacher ratio.

Compare the results in Table 4.5 to those in Table 4.6. The data in Table 4.6 reports the Pearson's correlation results using the same data. The latter results (i.e., direction and significance) for all three analyses are similar. There are a couple of things to consider as you look at these results together, however.

The Pearson's correlation (−.201) is very similar to the Spearman's correlation. This is usually the case when there are few "tied ranks," that is, when there are few values on a variable that are identical, and therefore that produce identical ranks for the Spearman's correlation analysis. Computationally the Spearman's correlation analysis calls for creating ranks in both variables, even (as in the present example) if

**Table 4.5    Correlation Results for TAGLIT Data**

| | | Correlations | | metro3_1 | studtch |
|---|---|---|---|---|---|
| Kendall's tau_b | metro3_1 | Correlation Coefficient | | 1.000 | −.163** |
| | | Sig. (2-tailed) | | | .000 |
| | | N | | 6615 | 6615 |
| | studtch | Correlation Coefficient | | −.163** | 1.000 |
| | | Sig. (2-tailed) | | .000 | |
| | | N | | 6615 | 10021 |
| Spearman's rho | metro3_1 | Correlation Coefficient | | 1.000 | −.210** |
| | | Sig. (2-tailed) | | | .000 |
| | | N | | 6615 | 6615 |
| | studtch | Correlation Coefficient | | −.210** | 1.000 |
| | | Sig. (2-tailed) | | .000 | |
| | | N | | 6615 | 10021 |

**Correlation is significant at the .01 level (2-tailed).

**Table 4.6    Pearson's r Results for TAGLIT Data**

| | Correlations | metro3_1 | studtch |
|---|---|---|---|
| metro3_1 | Pearson Correlation | 1 | −.201** |
| | Sig. (2-tailed) | | .000 |
| | N | 6615 | 6615 |
| studtch | Pearson Correlation | −.201** | 1 |
| | Sig. (2-tailed) | .000 | |
| | N | 6615 | 10021 |

**Correlation is significant at the .01 level (2-tailed).

one variable is ranked (metro3) and the other is interval (studtch). If the values to be ranked are dissimilar, then there will be fewer ties when the values are rank ordered. In practice, you can often use the Pearson's correlation for ranked data, but it is first important to examine the data to ensure that there are few tied ranks.

The Spearman's correlation (−.210) is higher than the Kendall's tau-b correlation (−.163). This reflects an important point in statistical analysis generally, that the most powerful result will follow from using the appropriate statistical test for the nature of the data in the analysis. In this case we could not use the correlation analysis for interval data (Pearson) since one of our variables was rank ordered (metro3). The Kendall's tau-b calculates the data differently from the Spearman's correlation, but is a powerful correlation procedure. In the current example, given the nature of the data (one interval variable, and a ranked variable based on interval values), the Spearman's correlation may be the best correlation to use.

## NONLINEAR CORRELATION

Often there are unobservable patterns to the data that we cannot see simply by examining the data with the naked eye. I am treating the correlation data in this book as if there are linear relationships among the variables. That is, I am assuming that the pattern of the data is that the values of the variables change together along a straight line, with the majority of the bivariate data points lying close to the line. Like the example in Figure 4.2, however, the data may be arrayed around a curved, rather than a straight line. See my comments in the "detail" section below for information about how to identify nonlinear correlations.

## "EXTENDING" CORRELATION TO INCLUDE ADDITIONAL VARIABLES

Thus far we have discussed correlation and regression procedures using two variables. By limiting our analyses to two variables, we have been able to understand how the values of the two variables relate to each other, and whether or not the values of one can predict the values of the other better than the mean of the outcome variable. However, we have also noted in places that the complexity of the real world goes far beyond the relationship between two variables. We now turn to statistical procedures that use the data from several variables simultaneously so that we can understand the various ways in which they all relate to one another. This ability can help in the quest for discovery, since new understandings often emerge from considering additional sources of information.

## CORRELATION—DETAIL FOR THE CURIOUS

### Computing Pearson's *r*

There are several methods for calculating *r* by hand. One common method is called the "*Z* score method" because it uses "standardized" scores. Standard scores are raw scores that are "transformed" into numbers that represent standard deviation amounts. There are many reasons to use these kinds of scores, but computing the Pearson's $r$ using $Z$ scores introduces an additional step (transforming raw scores to $Z$ scores) and uses a different formula.

Another method follows the "computational" formula, which calculates the correlation directly from the raw scores. There are several such formulas, so if you are interested, you can check introductory statistics textbooks to get a feel for how the formulas work. Regardless of which method you use ($Z$ score or computational), you should get the same answer because correlation indicates how the changes in one set of scores are associated with changes in the other set of scores (either raw scores or standardized scores). The computational formulas are bigger but use the data directly in raw score form. Using standard scores is quick, but when we get to the sections on regression, it is best to understand the raw score method so that you will be better prepared to interpret the results. Practically speaking, it doesn't matter which method we use because SPSS® will calculate the $r$ value for us. Still it is important to understand the different processes so that you can see how scores are used.

## Assumptions of Correlation

Most statistical procedures have certain requirements that must be met before they can be used effectively. These are akin to making sure that we use the "correct" data so that our method "fits" properly. Statistical procedures are like a bag of tools or instruments created for specific purposes. When the proper tool is selected for its chosen purpose, the best results will ensue. However, if a statistical tool is not appropriate for a given set of data or research situation, evaluators will not achieve optimal results. If you have need of a screwdriver, a pipe wrench will not do!

Correlation is often called "robust" because it will perform well even if certain of the assumptions discussed below are not strictly met. That is one of the many characteristics of this method that makes it so useful. But to have full confidence in our findings, we need to ensure that the assumptions are met.

Evaluation researchers use correlation heavily in their work. It is useful for associating different kinds of information, like achievement scores and class size, or work attitudes and units of production, for example. In most cases evaluators use correlation with an entire set of data with the purpose of understanding the strength of the connections between variables among all cases of the data available. In this situation, correlation is simply a way to *describe and discover* patterns without reference to the larger, unobserved, population from which the observed variables likely came.

Some evaluators, however, wish to treat correlation results as a reflection of population values, that is, how sample data reflect (or do not reflect) the population using an *inferential* process. In this event there are more careful rules about how the method can be used so as to have the most confidence in the results. The primary requirement is that the data be generated from a random sample where each individual has an equal chance of being chosen. This randomness allows the evaluator to assume that differences and errors are distributed in similar fashion to the population from which it was chosen, and therefore that the sample is not biased.

Evaluators should make sure to meet three assumptions to be reasonably assured of the best results for correlation procedures:

Assumption 1: Although Pearson's correlation works well with different kinds of data, it is best to use with *interval-level data* because we need to be able to assume that the data represent numbers that can be added, subtracted, multiplied, and divided meaningfully, with a specification about equal distances between data points.

This is one assumption that should be met regardless of whether you are using correlation descriptively or as a way to estimate unknown populations.[8] Evaluators who estimate population information from sample data need to view other assumptions especially carefully. By doing so, they can create "confidence limits" to the test results that indicate the extent to which they can be certain the sample data reflect the unknown population values. Statistical hypothesis tests create the protocols to ensure that we are taking the appropriate steps. (We will not discuss hypothesis testing here, so if you are unclear about what this means for correlation, you should refer to a good introductory statistics text.)

Under the strict conditions of hypothesis testing, the next two assumptions for correlation should be met:

Assumption 2: The sample data should be *randomly chosen* from the population.

---

[8]As we will see, there are evaluation analyses that call for the use of categorical data in the correlation-based process of regression. We will fully review this process when we discuss multiple regression.

This is a severe limitation to evaluation researchers because they often do not have the ability to identify an appropriate population and then randomly select individuals or cases for analysis. If they do have this population available to them, then they can make statistical inferences about the population's values based on the observed sample values, within the boundaries of confidence limits. If not, then they can only hope to describe accurately the patterns that might exist in the available data.

Assumption 3: Both variables should be *normally distributed*.

That is, each variable alone should be normally distributed and the values of both variables should be normally distributed at each of the levels of the other. In the first instance, you can use a graph (histogram) to see if both variables approximate a normal distribution. In the second case, say there are units of production; these then would be normally distributed at each of the levels of worker attitudes, as in the example described above.

## Nonlinear Correlation

Earlier I mentioned that some correlations can be based on nonlinear relationships. We could examine bivariate scatterplots to get a good idea of whether or not they appear to be linear, but there are a couple of other assessments we could use. The "Curve Estimation" menu in SPSS® is a procedure that will help you estimate the nature of the relationship between two variables. In the next chapter on regression, I have included a discussion of how to assess a curvilinear relationship through the SPSS® menus. At this point, we need to keep in mind that study variables may not be related to one another in linear fashion, and we should make every attempt to understand the nature of these relationships.

## DISCOVERY LEARNING

The main theme of this book is that, like the prism, the power of statistics can help the evaluator see hidden patterns in data when those patterns are exposed using the right methods. So let us review how the discussion thus far has aided in discovering hidden patterns. Simply by using a scatter diagram, we have been able to detect whether or not there may be a pattern to the data such that one variable is regularly linked to another. In the scatter diagram plotting math achievement and F/R, we observed that there was a "downward clustering" of the scores, which indicated that these two variables are related in inverse fashion. That is, as the value of one variable increases, the value of the other variable decreases. In the language of the data, schools with high percentages of children from poorer families have lower math achievement scores in the school as a whole. This is a substantial discovery, as educators, parents, and policy makers all are concerned about school math achievement levels not meeting a state's standards. There are a variety of possible answers to this question, but we have discovered one possible reason in our simple analysis. This discovery comes with a caveat, however: correlation is not causation. Just because two variables are related, even strongly related, that does not mean that one variable causes the other. In this instance family income is related to math achievement, but increases and decreases in achievement may be more directly caused by other factors. Or, family income and math achievement may both be related to other variables that may result in the correlation we have observed.

Using correlation methods adds a good bit of explanation for problems like this, but it is always difficult to make causal attributions, since there are so many other possible relationships "out there" that might be important. Other methods may help clarify these additional relationships, and we will take a look at them next.

Another discovery is the strength of the relationship between math achievement and family income at the school level. In our analysis the variables had a calculated correlation of −.62, which indicates that about 39% of the variance in school-level math achievement is accounted for by (low) family income of the students in the schools. This may not look like much on the surface, but one variable accounting for over a third of the variance in another is important. We want to remember this as we use additional methods to see whether this relationship holds.

Overall, if you are performing research, you will need to examine carefully what kind of data you have available, and make sure to choose the appropriate statistical procedure with your data. This will make your analyses more meaningful and accurate. *Here are some additional discoveries that we discussed in this section:*

- Correlation helps discover patterns in data.
- Correlation can illuminate the direction and strength of a relationship.
- Correlation helps to account for variance in the outcome variable.
- Visual graphs of data help expose patterns embedded in data.

## Terms and Concepts

**Aggregated data**   Data that are summed from one level and grouped at a different level. For example, aggregated achievement data are individual student-level scores that are summed together at the school level so that one (mean) achievement score represents the entire school's achievement level.

**Bivariate correlation**   A correlation between two variables.

**Categorical data**   Non-interval data that are in a limited number of discrete groups. Typically these data represent "types" of information like gender or demographic qualities like "urban."

**Coefficient of determination**   The square of the correlation coefficient that represents the amount of variance in one variable "accounted for" by the other variable.

**Correlation**   The extent to which changes in the values of one variable are related to changes in the values of another variable. Positive correlations are those in which increasingly higher values of one variable are related to increasingly higher values of another. Negative (or inverse) correlations are those in which increasingly higher values of one variable are related to increasingly lower values of another. A "zero" correlation is one in which there is no relationship between the values of one variable and the values of another.

**Curvilinear relationship**   A "nonlinear" relationship between the values of two variables whereby the values of one variable can be low at both low and high values of the second. The resulting pattern is a curved series of dots.

**Ecological fallacy**   This is a mistake in an interpretation whereby the evaluator draws conclusions from one level of data as the analyses are performed on a different level of data, for example, uses school-level achievement scores to arrive at conclusions about individual student achievement.

**Factor analysis**   A statistical technique that allows a researcher to discover relationships among "clusters" of items on an instrument that are highly related to one another but have lower relationships to other clusters. The assumption is that each item reflects a certain analytical meaning on the part of the respondent that can be detected by its relationship to other items.

**Nonparametric correlation procedures**   Measures of correlation not based on interval data or their referent parametric distributions. Typically these correlation measures are used with nominal and/or ordinal data. Some examples used in SPSS® are Kendall's tau, the phi coefficient, and the contingency coefficient.

**Outlier scores**  Scores in an analysis that represent extreme values. Such cases may result in analyses that are inordinately affected by the distance of an extreme score from the remaining scores.

**Pearson's correlation ($r$)**  An association procedure named after Karl Pearson in which interval level variables are measured against other interval level variables (also known as "Pearson's product-moment correlation").

**Restricted range**  A difficulty with correlation procedures in which only a portion of the values of one variable are correlated with the values of another. The result is a distorted view of the relationship, since much of the information is not included in the analysis.

**Scatter diagram**  A graph in which values on two variables are represented on a matrix where one variable is designed as the predictor variable ($X$) and the other variable is designed as the outcome ($Y$). Each case in the database is located by a dot that contains the simultaneous values of $X$ and $Y$. The resulting graph displays an arrangement of dots that may fall into a recognizable pattern or a completely random array.

**Spearman's rank order correlation**  A correlation procedure used when one or both variables are ordinal in nature ("greater than" or "less than").

## Real World Lab—Correlation

Because the aim of this book is to help evaluators use statistics to discover patterns in data, we will use actual data from evaluation projects for the processes we discuss. The following is the first practical application of using actual data with correlation procedures. I will describe the dataset and provide directions for how to proceed. The answers to this and the following practical application questions are posted in the Practical Application Analyses section at the end of the book.

## Description of the Data

The data set presented in the sidebar is a sample of the larger data set of classroom observation results aggregated to the school level. The data represent a series of observations from classrooms of different subjects (math, eng., soc. studies, etc.) in different schools aggregated to the school level. The focus is research on the nature of teaching and learning in various subjects in elementary and middle schools in the five essential components of the STAR Classroom Observation Protocol™. The four variables represent information on 32 schools during the 2005 to 2006 school year, and include achievement results in reading ("readstd"—percentage of the students at the school who met the achievement test standard in reading) and math ("mathstd"—percentage of the students at the school who met the achievement test standard in math), as well as a measure of the family income of students in the school ("percfrl"—the percentage of the students at the school eligible for free or reduced price meals) and the observation score (STAR_mean).

The STAR Classroom Observation Protocol™ is a research-based instrument designed to measure the degree to which Powerful Teaching

| mathstd | readstd | percfrl | STAR_mean |
| --- | --- | --- | --- |
| 80.60 | 94.00 | 28.25 | 2.29 |
| 63.30 | 83.30 | 29.35 | 2.52 |
| 32.90 | 67.10 | 97.63 | 2.15 |
| 60.70 | 89.30 | 69.40 | 2.93 |
| 40.00 | 83.10 | 40.72 | 3.18 |
| 34.20 | 76.30 | 39.66 | 2.33 |
| 59.70 | 80.50 | 94.00 | 2.78 |
| 44.10 | 75.00 | 19.27 | 2.77 |
| 44.60 | 95.20 | 96.78 | 3.30 |
| 36.70 | 69.40 | 82.92 | 2.64 |
| 39.80 | 76.10 | 66.34 | 2.94 |
| 15.20 | 54.30 | 93.74 | 2.92 |
| 48.90 | 76.10 | 79.27 | 2.72 |
| 30.60 | 82.10 |  | 2.60 |
| 32.50 | 79.20 | 86.15 | 2.36 |
| 29.70 | 73.00 | 88.09 | 2.74 |
| 31.30 | 49.60 | 90.05 | 2.63 |
| 56.40 | 77.70 |  | 2.30 |
| 40.50 | 62.20 | 46.68 | 2.81 |
| 60.20 | 87.40 | 21.23 | 2.59 |
| 26.70 | 69.90 | 88.59 | 2.35 |
| 22.90 | 52.00 | 98.68 | 2.33 |
| 44.60 | 82.10 | 32.10 | 2.14 |
| 17.80 | 43.00 | 60.87 | 2.48 |
| 20.70 | 44.60 | 44.47 | 2.04 |
| 50.00 | 50.80 | 89.21 | 2.11 |
| 15.90 | 38.40 | 84.76 | 2.56 |
| 33.60 | 54.40 | 57.82 | 2.41 |

| mathstd | readstd | percfrl | STAR_mean |
|---------|---------|---------|-----------|
| 33.80 | 57.30 | 57.82 | 2.36 |
| 20.60 | 35.90 | 93.35 | 2.12 |
| 20.90 | 25.30 | 94.74 | 2.41 |
| 42.00 | 60.30 | 68.32 | 2.74 |

and Learning™ is present during a classroom observation.[9] As part of the design of the STAR™ protocol, only the most significant and basic indicators are used to determine the presence of Powerful Teaching and Learning™. Thus the STAR™ protocol allows for ease of use with any classroom observation and aligns with the educational improvement goals and standards for effective instruction. The STAR™ protocol helps participants view Powerful Teaching and Learning™ through the lens of 5 essential components and 15 indicators. The goal of this data collection is to determine the extent to which general instructional practices throughout the school align with Powerful Teaching and Learning™.

## Evaluation Questions

The objective here is to conduct a correlation analysis among the variables in the data set using the SPSS® menus as described in the previous chapter. You should examine the correlation matrix and respond to these questions:

1. Are any of the correlations significant?
2. What is the coefficient of determination, or effect size, of the relationship between reading achievement and the STAR_mean?

Then create a scatter diagram of reading achievement as the outcome variable and STAR_mean as the predictor variable. Describe the resulting pattern of the data. Last, express the correlation results in the language of the research problem.

[9]These data are used with the permission of The BERC Group, Inc.

# 5

# REGRESSION

Regression has many uses in research, including the ability to *predict* values of an outcome variable, as well as to provide *explanation* and testing models of data. As with correlation, upon which regression is based, the use of regression depends a good deal on what assumptions the evaluator makes about the nature of the method and what steps are needed for identifying and collecting data.

Much of the time those who are engaged in evaluation research use regression analyses on a set of available data taken from a project and wish simply to understand the patterns of relationships. For example, a school district leader might hire an evaluator to analyze a set of survey responses from the teachers in the district about their teaching style and effectiveness.

In this situation the evaluator can use regression results to describe the relationships among the variables with a focus on discovering patterns of data, but without a theoretical assumption about which variables should relate to which others or in what ways. The primary use for regression in these instances would be to understand the patterns in the data at hand, and perhaps to *predict* values of an outcome variable with known predictor values. *Predicting values does not rely on theoretical assumptions.*

*The Program Evaluation Prism: Using Statistical Methods to Discover Patterns,*
by Martin Lee Abbott
Copyright © 2010 John Wiley & Sons, Inc.

In other situations an evaluator can proceed by conducting the data analysis according to theoretically driven or a priori assumptions. That is, the evaluator might interpret the findings through a rationale of how the variables should relate to one another according to a prevailing theoretical model. In this case the evaluator could use the findings for *explanation*. The evaluator might then make statements about the amount of the variance in the outcome variable being explained by the set of predictor variables. *Explanation requires a more rigorous set of assumptions than prediction.*

Of course, in both of these instances the evaluator would need to take great care in identifying and gathering the data for analysis. In many other cases evaluation researchers can make only limited claims about relationships among variables or about relevant theories because they might only have access to limited data (e.g., one school district). Therefore the evaluation researcher might be able to get a good description of the responses of the teachers in that district (making such assumptions as 100% return rate, etc.) but would not be able to generalize the findings beyond the school district. The target district may not be characteristic of other school districts in the same region or state.

Evaluation researchers can make more extensive theoretical comments and generalize the findings further if they use probability-based sampling methods. In this example that might mean selecting teachers randomly from several districts (also chosen randomly). Most often, however, evaluators are limited in their ability to perform probability-based sampling and must be content with a limited set of data.

In the following sections on regression, I will speak of predictor variables "accounting for variance" in the outcome. In making this statement, my aim will be to make a distinction between situations where explanation is the goal, and where we are only trying to describe the patterns of relationships and perhaps to make appropriate predictions.


## THE REGRESSION LINE—LINE OF "BEST FIT"

In the previous chapter we examined several scatter diagrams in an attempt to describe the features of correlation. We can extend that information by introducing a way to use these scatter diagrams to help illustrate how evaluators can use knowledge of the correlation to better predict values. Scatter diagrams like the one in Figure 5.1 show that the scores have a pattern we identified earlier as being a positive correlation—where the scores generally group together and fall in an upward direction.

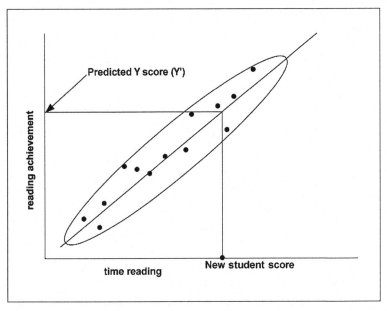

**Figure 5.1** Line of best fit for the relationship between time reading and reading achievement.

If we were to draw a line through our scores, such that the sum of the squared distances between each point and the line was as small as possible, we would have a "line of best fit." This line is the "regression line," and we can use this line to help us see how we can predict values of an outcome variable—"$Y$"—if we have a significant correlation.

Suppose that the data in the figure represented students at a particular school, and that we concluded that the correlation between reading achievement and time spent reading was statistically significant. In this case there would be a pattern to the scores, and they would fall closer together the stronger the correlation. As the figure shows, the scores are all clustered around the line. We can use the regression line to help us predict a reading achievement score for a new student to class without administering a test if we know how much the new student reads. We simply draw a vertical line from his time reading score to the regression line, and then extend our line horizontally to the $y$ axis, which indicates reading achievement. This is the predicted value of $Y$ (designated as $Y'$). If there were no significant correlation to begin with we would not be able to do this, since the regression line could not be drawn with accuracy, as shown in Figure 5.2.

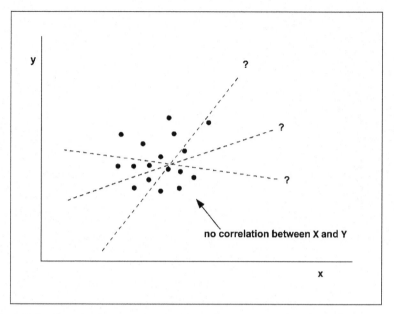

**Figure 5.2**   Scatter diagram showing no correlation between variables *X* and *Y*.

As Figure 5.2 shows, when there is no discernable pattern to the data, we cannot be sure where the regression line should be placed. Even if we arbitrarily place the regression line, or used different methods for fixing it, the data would not be tightly clustered to the line, and the resulting prediction of a *Y* score would be highly inaccurate. Recall what we saw earlier, the stronger the correlation, the closer all the scores are to the line. The closer they are to the line, the more "accurate" our prediction of *Y*, as shown in Figure 5.3.

The data in the left panel indicate that our predicted value (*Y′*) could have taken a wide range of values, if we think of the scatter around the line as being a kind of distribution of scores around a mean, with the mean represented by the regression line. So the prediction in the left panel could take values from the first dotted line on the lower side of the spread of scores (designated by *a*) to the topside of the spread (designated by *b*). The right panel shows how the prediction improves when the scores are closer to the regression line. When the correlation is stronger, the prediction is better (i.e., the distance from *a* to *b* is smaller).

If the original correlation is weak or nonexistent, we would be unable to draw a regression line. In that situation the mean of *Y* would be the best prediction for a value of *Y*, since our predictions could take any

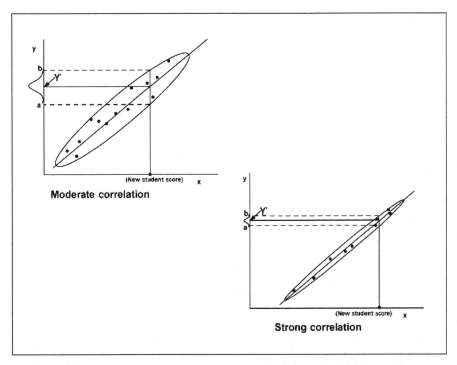

**Figure 5.3**  Prediction accuracy as a function of the strength of correlation.

range of values, and the mean of the distribution is the one that is the most characteristic of all the other scores. This would be equivalent to just saying, "forget $X$—if we don't have a correlation, it is not helpful—therefore if I want to predict any value of $Y$, I will always choose the mean of $Y$."

We need to have some basic information about the regression line before we can use it for prediction. We have discussed the first criterion—the extent to which the scatter is close to the line. Second, we need to understand the nature of the scatter, which affects the slope of the line, also discussed earlier. Last, we need to know where the line crosses the $y$ axis. Technically this is the "intercept," or the value of $Y$ when $X$ is zero. If we know these things, we can construct our regression line.

## THE REGRESSION FORMULA

I will not spend much time in this book discussing the mathematical derivation of the formulas we encounter. In the case of the regression formula, it is important to know just a few basic facts. First, what do each of the elements represent, and second, how can the evaluator interpret the findings.

The regression formula often appears in different forms, but the following is fairly standard:[10]

$$\hat{Y} = bX + a$$

"$\hat{Y}$" (or "$Y$ hat") is the *predicted value of y*. This value is sometimes symbolized as $Y'$.

Here $a$ is the *Y-intercept*, or the place where the regression line crosses the y axis. It is listed in many formulas and statistical analysis programs as $b_0$, since it refers to the estimated value of $Y$ when $X = 0$ (or, in the case of multiple regression, when the predictors = 0). $b$ is the *slope* of the regression line. It is also a regression estimate, with a subscript indicating a reference to a specific variable. When there is only one predictor variable, there may be no subscript, or it might appear as $b_1$ to refer to the change in $Y$ with a one unit change in the first predictor variable $X_1$. If there are multiple predictors, the estimates for subsequent predictors would be designated $b_2$, $b_3$, for $X_2$ and $X_3$, and so on, where the estimate would indicate changes in $Y$ with unit changes in the appropriate predictor, when the other predictors are held constant.

Both the $a$ and the $b$ have separate formulas for their calculation. But we will see below how to recognize these in the SPSS® output. Just remember that we are predicting values of $Y$ from using $X$, based on a significant correlation between $X$ and $Y$. As you might imagine, our predictions of $Y$ will likely not be equal to the actual value of $Y$, if we could know the actual value of $Y$. They will be close, but there will almost always be a discrepancy. This discrepancy is the *residual error*, and is very important in some later analyses.

On the basis of how we use the regression equation, according to the discussion above, the following is a useful designation for the regression formula:

$$\hat{Y} = b_0 + b_1 X_1$$

We should also note at this point that there are several ways of testing regression hypotheses. You can test the hypothesis that the true slope of the regression line = 0. A rejection of the null hypothesis would indicate that the true slope of the line has a value other than 0 and

---

[10] You may recognize this from an old algebra class as a variant of the formula $y = mx + b$.

can be helpful in prediction. You can perform an overall "*F* test" to determine if the slope is significantly different than 0, and SPSS® also provides a "*t* test" for individual predictor variables.

There are several concepts and measures important for understanding how to create and interpret regression findings. I will discuss three such measures in the next sections and then discuss related topics in the "Detail—For the Curious" section below.

## STANDARD ERROR OF ESTIMATE

The *standard error of estimate*, SE(est), is much like the standard error of the means that is discussed in introductory statistics classes. We are using our calculated values from the sample to create measures that will help us see how accurate our estimates are of the outcome variable *Y*. This population estimate is a measure of the variation of the data points around the regression line. Stated differently, it represents the distribution of the differences between our *estimated* values and the *actual* values of the data points. Look at Figure 5.3 to see how there would be difference in SE(est) figures. The SE(est) for the result in the top (left) panel would be larger than that of the result in the bottom (right) panel, since the estimated values are likely to fall further from the actual values of *Y*.

The value of the SE(est) can range between 0 (no variation) and the standard deviation of *Y*. You can see how this might be true if you look at the formula:

$$SE(est) = SD_Y \sqrt{1 - r^2}$$

1. If $r = 0$ (or no significant correlation), then the result is multiplying $SD_Y$ by 1 ( $\sqrt{1}$ ). In this case you would get $SD_Y$. This restates what we noted earlier.
2. If $r$ is perfect (+/–1.00), the result is multiplying $SD_Y$ by 0, which would equal 0.

The formula above is helpful to understand the principle of the SE(est). In practice, formulas are available that take into account the sample size of the data, especially if you are working with smaller databases (e.g., 30 or less) and if you use the actual or estimated standard deviation. In practice, however, you will use the results provided by statistical programs such as SPSS®, so the key issue is what SE(est) represents, rather than how it is calculated.

## CONFIDENCE INTERVAL

*Confidence intervals* are boundaries within which the true population values are likely to fall, given the information derived from sample values. With prediction, we are estimating the true population value, in this case the true $Y$ value. We can use the SE(est) to calculate the interval within which we can expect the true (parameter) value of $Y$ to be. To do this, we can use a formula familiar to students of introductory statistics for estimating the population mean within a certain confidence interval using SE(mean). We can use the $Z$ distribution to help us establish the limits of this interval (as long as we have a large enough sample).

Using the formula ($CI_{.95} = z[SE(est)] + Y'$), we can see where the actual value of $Y$ might lie within certain bounds of confidence. In this formula we would use +/–1.96 as the $Z$ critical value. For example, if you wanted to establish a .95 confidence interval around a predicted $Y$ value of 1.36, and the SE(est) = .64, using the formula would result in two figures (using "+" and "–" values of 1.96) of .11 and 2.61. Therefore we would have a .95 confidence that the actual value of $Y$ would be between .11 and 2.61.

Confidence intervals are very important for evaluators because they provide a measure of understanding how sample values relate to the unmeasured populations from which they supposedly come. In the calculation above we demonstrated how specific predicted values could have confidence intervals. All regression estimates can have confidence intervals that serve as indications of the extent to which we can be certain of our findings. As we will see, SPSS® output always provides the option for creating confidence intervals, and we will show how these apply to specific examples of data.

## RESIDUALS

As I mentioned earlier, *residuals* are measures of error. They measure the difference between predicted scores and actual scores. We will look at residuals in more detail later, but for now it is important to understand how they are created, and what they can tell us about our data estimates.

If you go back to our previous discussion of predicted values, you will recall that we created predictions based on a significant correlation between a predictor and an outcome variable. The predicted value of a particular score thus contains a lot of information. Its value is partially determined by the strength of the correlation, but also by unknown influences. Therefore a predicted value can be broken down into these "known" (through correlation) and "unknown" portions.

When we look at all the residual scores together with their actual scores from a set of data, we can understand these relationships more clearly. The following is an important formula that describes the relationships among these portions:[11]

$$\sum (Y - \bar{Y})^2 = \sum (\hat{Y} - \bar{Y})^2 + \sum (Y - \hat{Y})^2$$

There are essentially three measures of variation in this formula based on the *sum of squares* calculation.[12] As you may recall from introductory statistics, the sum of squares is a calculation of squared differences between individual scores and their means. Thus, the formula above specifies the portions or components of variation embedded into the overall sums of squares of Y, our outcome variable. This is the portion to the left of the equals sign.

$$\sum (Y - \bar{Y})^2 = \text{sum of squares of } Y \text{ scores, or simply, SS}(Y)$$

The next component (immediately to the right of the "=" sign) represents the "known" portion of our outcome variable in that it reflects the information gained from the correlation of Y with one or more predictors. This portion of the overall sum of squares of Y is called the *sum of squares of regression* because it captures this contribution of the predictor(s) to the outcome. As you can see, it is simply the sum of squares of differences between predicted scores and the mean of Y. The stronger the relationship between predictor(s) and Y, the larger this value becomes relative to the other portion. Think of the strength of the relationship "driving" the predicted value further away from the mean, closer to the actual value of Y.

$$\sum (\hat{Y} - \bar{Y})^2 = \text{sum of squares of regression, or SS(reg)}$$

The last portion represents the part of the sum of squares of Y whose origin we cannot identify. It can therefore be called an "error" since its

---

[11] In this formula, $\bar{Y}$ refers to the mean of Y, and $\hat{Y}$ is the predicted value of Y.

[12] The sum of squares calculation is not the variance, which is obtained by dividing the sum of squares by $N$ or $N - 1$, and so forth. Rather, it is the accumulated squared deviations of scores from their mean. You can think of it as a rough index of variation in that the larger the number, the greater is the difference of the scores from their means.

value is oriented outside the correlation between our predictor(s) and our outcome variables. In fact it is often called the "sum of squares of error." As you can see, this portion is represented by the sum of squared differences between actual values of $Y$ and predicted values of Y. It is therefore commonly called the *residual sum of squares*. Technically it is a combination of a random error and an unexplained variation due to additional factors not in the analysis. With additional predictors, as is the case of multiple regression, we might be able to understand or explain more of this variation.

$$\sum \left(Y - \hat{Y}\right)^2 = \text{sum of squares of residual, or SS(res)}$$

If you think about these portions of the overall variation in the sums of squares of $Y$, you will see that the greater the value of the SS(reg), relative to the SS(res), the stronger is the prediction (and explanation) of the outcome $Y$ from the predictors. You may even recognize that the coefficient of determination, or $r^2$, can be calculated from these portions. Consider the following formula:

$$R^2 = \frac{\text{SS(reg)}}{\text{SS}(Y)}$$

Figure 5.4 shows how these relationships would appear on a graph. To the graph I have added the proportions of variation as $a$, $b$, and $c$ to illustrate the relationships among proportions within a single case. The portion $a$ is the total measure of variation that exists between an actual score and the mean of $Y$. The portion $b$ is the distance between the mean of $Y$ and the predicted $Y$ score—this is the distance representing how far the correlation "moves" the predicted score toward the actual score. In many ways this represents the effect of the correlation, since a stronger correlation will result in less distance between actual and predicted scores. Proportion $c$ is the residual, or the distance from the predicted to the actual score, whose origin is not identified in this analysis.

From these proportions you can see that $R^2$ expresses the relationship between the total variation ($a$) and part $b$—as the regression variation increases, it represents a larger proportion of the overall variation. This relationship represents the percent of total variation in $Y$ accounted for by its relationship to the predictor variable $X$.

All we have done here is to show that the amount of variance accounted for (represented by $R^2$, the coefficient of determination) in the relationship between the outcome and predictor(s) equals the

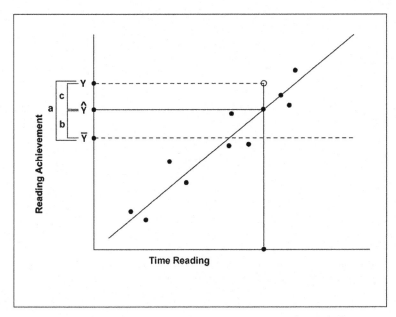

**Figure 5.4**    Components of the sum of square values.

**Table 5.1    Example Data Showing the Proportion of Sums of Squares**

| $X$ | $Y$ | Predicted $Y$ | SS($y$) | SS(reg) | SS(res) |
|---|---|---|---|---|---|
| 2 | 1 | 1.15 | 23.04 | 21.62 | .02 |
| 4 | 5 | 4.25 | .64 | 2.40 | .56 |
| 5 | 4 | 5.80 | 3.24 | .00 | 3.24 |
| 6 | 9 | 7.35 | 10.24 | 2.40 | 2.72 |
| 8 | 10 | 10.45 | 17.64 | 21.62 | .20 |
| | | | 54.8 | 48.05 | 6.75 |

proportion of the total sum of squares of $Y$ represented by regression, or SS(reg).

All these portions may best be seen through an example. Consider the following data in Table 5.1.[13] The first two columns are the raw data, and the remaining columns are the regression results. Notice that the SS(reg) is large relative to SS(res), which indicates that the regression portion is a greater part of the overall sum of squares of $Y$. This is a clue that the $R^2$ value will also be large. You can see the overall equation represented in the following ways:

---

[13]We will discuss below how to create predicted and residual values in SPSS®.

$$\sum(Y-\bar{Y})^2 = \sum(\hat{Y}-\bar{Y})^2 + \sum(Y-\hat{Y})^2$$
$$SS(Y) = SS(reg) + SS(res)$$
$$54.8 = 48.05 + 6.75$$

Now you can see the $R^2$ formula through actual values from the table:

$$R^2 = \frac{SS(reg)}{SS(Y)}$$

$$R^2 = \frac{48.05}{54.80} = .877$$

The are other uses for residuals, but this discussion can help you see how residuals can be used to interpret the relationship between outcome and predictor(s). We will discuss some of the other aspects of residuals in later sections.

## REGRESSION EXAMPLE WITH ACHIEVEMENT DATA

Let us take a concrete example with the data we examined earlier, the relationship between math achievement and F/R at the school level. To make the graphs simpler, I took a random sample of cases from the original database, as shown later in Figure 5.9. You will note that the correlation for this sample is slightly higher than that calculated on the entire database (–.62), but that is the nature of samples. We will use this sample database only because it has fewer cases and therefore makes some of our operations below that deal with individual data points a bit clearer. We will return to the overall database for our primary analyses.

Figure 5.5 shows the steps using SPSS® to run the bivariate regression. At the opening screen, choose "Analyze" and then "Regression"—you can choose "Linear" when that submenu appears. Making this choice will result in the submenu shown in Figure 5.6, which allows you to specify which variables you will use in your analysis and what specifications you would like to include.

At this point you would use the arrow keys to select which variable you would like to specify as the "Dependent" variable and place it in the window to the right of the arrow. Similarly you would use the "Independent(s)" arrow to select the independent variable for

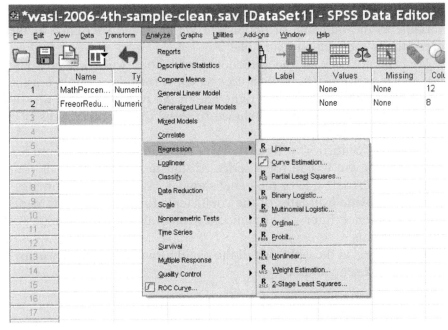

**Figure 5.5**    SPSS® screen showing menu for linear regression.

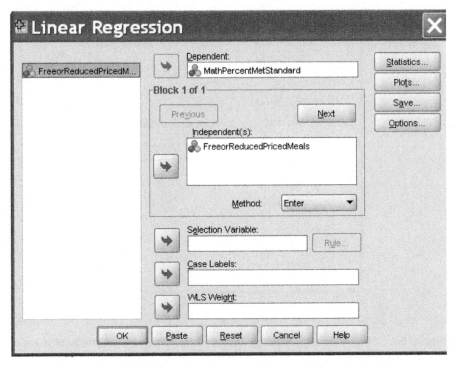

**Figure 5.6**    SPSS® screen showing submenu for linear regression.

the analysis. You will see that I have specified the math achievement variable as the outcome or dependent variable, and FRLunch as the predictor or independent variable.

We will not use all the selection possibilities with this bivariate regression example. The key choices to select for this example are the series of buttons in the upper right side of the panel. For now we will only focus on the "Statistics" choice to show the overall results of a bivariate regression analysis.

Choosing "Statistics" will result in the submenu shown in Figure 5.7, which allows you to call for a series of statistical procedures in your analysis. As the left side of the panel shows, I have chosen "Estimates" (this will provide estimates of the coefficients for slope and intercept) and "Confidence Intervals" (confidence intervals for the estimates of the coefficients.

The choices on the right side of the panel provide the primary analyses of the chosen variables:

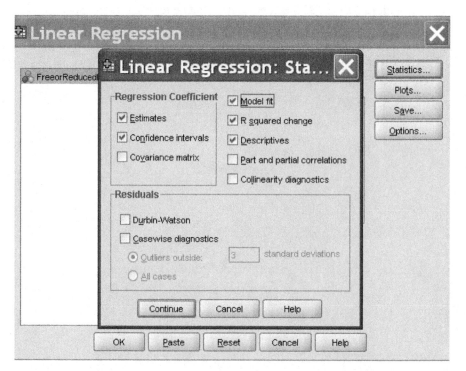

**Figure 5.7** SPSS® screen showing submenu for specifying elements of a linear regression analysis.

- "Model fit" provides the ANOVA table, which provides the results for the omnibus test.
- "$R$ squared change" provides information about how the $R^2$ for the model changes as additional independent variables are added to the analysis. In the current example we will have only one independent variable, so the importance of this information will be explored in later analyses.
- "Descriptives" provides means and standard deviations for the variables chosen for analysis, as well as a correlation matrix.
- "Part and partial correlations" will be explored in much greater depth later.
- "Collinearity diagnostics" provides information helpful for determining the extent to which we meet the assumptions for this test and will be explored in detail below.

There are choices possible if we wish to examine the "Residuals" for the analysis, but we will not explore this feature at this time.

## The Results of the Analysis

The *SPSS® output* for the regression analysis using the sample database consists of several parts. We will show each of the parts and comment on the interpretation of the results. The first part of the output is shown in Table 5.2. This part of the output shows the descriptive statistics (means and standard deviations), and the correlation between the variables in the analysis, shown in Table 5.3, is the result of choosing the "Descriptives" button.

The output in Table 5.3 shows the correlation between math achievement and F/R in the sample of schools we choose for illustration. As shown, the correlation coefficient for the 89 schools was –.643. Another part of the SPSS® output is shown in Table 5.4.

The ANOVA results represent the "omnibus" findings, or the overall test of whether or not the model is statistically significant. (This output is created by choosing the "Model fit" button.) In this case the $F$ test

**Table 5.2  Descriptive Output for Regression Analysis**

| | Mean | Std. Deviation | N |
|---|---|---|---|
| | Descriptive Statistics | | |
| MathPercentMetStandard | 59.7663 | 16.40722 | 89 |
| FreeorReducedPricedMeals | 41.4928 | 24.04742 | 89 |

Table 5.3 Correlation Matrix for Regression Analysis

| Correlations | | | |
|---|---|---|---|
| | | MathPercent MetStandard | FreeorReduced PricedMeals |
| Pearson Correlation | MathPercentMetStandard | 1.000 | −.643 |
| | FreeorReducedPricedMeals | −.643 | 1.000 |
| Sig. (1-tailed) | MathPercentMetStandard | | .000 |
| | FreeorReducedPricedMeals | .000 | |
| N | MathPercentMetStandard | 89 | 89 |
| | FreeorReducedPricedMeals | 89 | 89 |

Table 5.4 Omnibus (ANOVA) Output for Regression Analysis

| ANOVA[b] | | | | | |
|---|---|---|---|---|---|
| Model | Sum of Squares | df | Mean Square | F | Sig. |
| 1 Regression | 9,781.984 | 1 | 9,781.984 | 61.193 | .000[a] |
| Residual | 13,907.334 | 87 | 159.854 | | |
| Total | 23,689.319 | 88 | | | |

a. Predictors: (Constant), FreeorReducedPricedMeals.
b. Dependent Variable: MathPercentMetStandard.

Table 5.5 Model Summary Output for Regression Analysis

| Model Summary | | | | | | | | |
|---|---|---|---|---|---|---|---|---|
| | | | | | Change Statistics | | | |
| Model | R | R Square | Adjusted R Square | Std. Error of the Estimate | R Square Change | F Change | df1 | df2 | Sig. F Change |
| 1 | .643[a] | .413 | .406 | 12.64335 | .413 | 61.193 | 1 | 87 | .000 |

a. Predictors: (Constant), FreeorReducedPricedMeals.

(61.193) is statistically significant ($p < .001$). Therefore the overall model of FRLunch predicting math achievement for this group of schools is statistically significant.

Table 5.5 derived from SPSS® output provides the next part of the regression analysis resulting from choosing "$R$ squared change" from the menu. This part of the output contains several important findings for the analysis. The "Model Summary" shows the $R$, or correlation between the output and input variable. Since we have only one predictor, the $R$ is equal to the Pearson's correlation ($r$) reported in the correlation matrix shown in Table 5.3.

The coefficient of determination ($R^2$) of .413 is shown next. This is the variance in the outcome variable accounted for by the predictor variable(s) in the analysis. Since the ANOVA table reports the "sum of squares" for both the regression and the residual, you can confirm how much of the variance in the output variable is explained by the predictor by dividing the "regression sum of squares" by the "total sum of squares." (See the previous discussion and formula.) This allows you to see the proportion of the variance in the outcome variable (math achievement) accounted for by the correlation with the predictor variable (F/R). In this example, F/R accounts for .413 or 41% (from 9781.984/23,689.319) of the variation in math achievement.

The adjusted $R^2$ is a value showing the likelihood of the $R^2$ value that would be obtained with all the data from the actual population in the analysis. A correlation or $R$ value based on a limited sample size, with only one predictor, may result in a larger correlation than if all the values of the population were involved. The adjustment to $R$ takes into account sample size and number of predictors to provide a value that would more likely reflect a population-based analysis. This is one reason to have as large a sample size as possible. In this example there is a large discrepancy between $R^2$ and "adjusted $R^2$," which is mostly influenced by a small sample size ($N = 89$). The standard error of the estimate is created using the adjusted $R^2$, so it will be slightly different from the one created using the unadjusted $R^2$.

We will discuss the "change statistics" in a later chapter. Overall, these SPSS® output tables show how the $R^2$ and other indicators change as a result of adding variables to the analysis. In this case there is only one independent variable, so the change statistics simply reflect the fact that our correlation changed our explanation from zero to an $R^2$ of .413.

The last part of the SPSS® output for this analysis is the "Coefficients" results shown in Table 5.6. We will provide additional, detailed, explanations of these types of results in later sections of the book. However,

**Table 5.6  Coefficients Output for Regression Analysis**

| | Coefficients[a] | | | | |
|---|---|---|---|---|---|
| | Unstandardized Coefficients | | Standardized Coefficients | | |
| Model | B | Std. Error | Beta | t | Sig. |
| 1  (Constant) | 77.958 | 2.684 | | 29.045 | .000 |
| FreeorReducedPricedMeals | −.438 | .056 | −.643 | −7.823 | .000 |

a. Dependent Variable: MathPercentMetStandard.

I want to point out some important features of the regression output here. Table 5.6 shows the coefficients for creating the regression equation. There are values for the slope of the line and the intercept. Further notice that there is a "*t* test" of the predictor variable ($t = -7.823$). This value indicates the probability that the slope of the regression line resulting from the relationship between math achievement and FRLunch is 0, or flat. As you can see, the significance of the *t* value is .000, which indicates only a very small probability that there is a zero slope resulting from chance alone. That is, the slope for the regression line based on this predictor is statistically significant.

With one predictor, the omnibus results and the results for the individual predictor are closely related because only one predictor is responsible for the omnibus findings. You can see this relationship by squaring the *t* value, which will equal the *F* value in the omnibus table ($-7.823^2 = 61.199$).

The overall regression equation can be obtained by using the estimated coefficients in this panel of Table 5.6. As shown in the "Coefficients" panel of the table, the unstandardized beta is −.438, which indicates that math achievement decreases by .438 (not quite one half of a percentage point) for every 1% increase in F/R. The "Constant" value shown is the *Y* intercept, that is, the value of *Y* when *X* = 0. Recall that the equation for the regression line is, $Y' = bX + a$. Using the data reported in the SPSS® output panel, the equation for the regression line in this example is therefore

$$Y' = (-.438)X + 77.958$$

You can use this equation to find predicted values of math achievement when you know values of F/R. It may not be common practice to create predicted values in research situations involving individual respondent data, unless the express purpose of the research is to create a predictive model that can be used for specific purposes by those who sponsor the research. A predicted value is more commonly used with aggregated data, since not all data points are always available. For example, if we are interested in identifying a school for further evaluation but have limited data, we might use the regression formula to predict a value on the outcome (in this case math achievement) based on the predictor (in this case FRLunch percentage).

When using unstandardized coefficients, as in the equation above, the original scale of the variables is retained so that we can make an interpretation such as the one above. If we use the *standardized* coefficient, we need to remember that the predicted $Y$ scores are in "standard deviation language" (or $Z$ scores). This would result in the regression equation of $Z_Y = \text{Beta}(X)$, or in this example,

$$Z_{Y'} = (-.643)Z_X$$

In this case both variables were transformed by the statistical program to standard scores so that they have the same scale.[14] As shown in the output in Table 5.6, the beta is $-.643$, indicating that when F/R reduces by one standard score, math achievement reduces by .643 standard scores.

When we discuss multiple regression, the SPSS® results will report separate coefficients for each independent variable, so we will see which predictors are considered significant by looking at the $t$-test information. We can also use the estimated coefficients to create the regression equation.

## The Graph of the Results

As shown in Figure 5.8, a regression line is calculated for these data and included in the graph. Following the scatter of the scores, the line shows an inverse correlation where the pattern of cases extend from upper left to lower right. The bold horizontal and vertical lines on the graph illustrate how the regression line can be useful for prediction. In this particular case, we chose an FRLunch value of 40 to see what prediction of math achievement would result. As shown, the predicted value of $Y$ is where the bold horizontal line intersects the $y$ axis (just above 60). We can use the regression equation that we derived above to provide a specific value of the predicted math achievement value. Using $X = 40\%$ in the regression equation, we have the resulting predicted math achievement value as 60.438%:

$$Y' = (-.438)X + 77.958$$

$$Y' = (-.438)(40) + 77.958$$

$$Y' = (-.17.52) + 77.958$$

$$Y' = 60.438$$

---

[14] Recall that standard scores (or $Z$ scores) represent scores from a distribution with a mean of 0 and a standard deviation of 1.

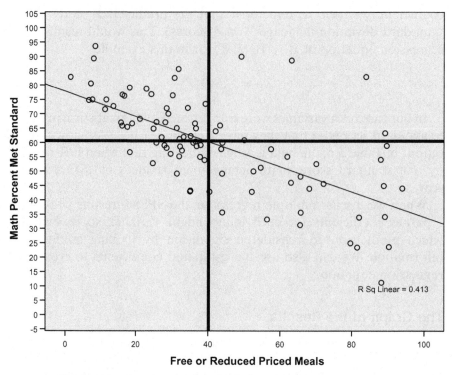

**Figure 5.8** Creating predicted values using the regression line.

## Standard Error of the Estimate

Figure 5.8 illustrates other features of regression analysis that we discussed earlier, in particular, the standard error of the estimate and the confidence intervals for our prediction. You will notice that the $X$ (FRLunch) value of 40 is a value that may not exist in the database. This may be one of the cases where we can use the significant regression equation to help us predict an unknown outcome variable when we have only a known predictor. For example, we may not know the math achievement score for a certain school for which we only have a FRLunch value. In this case we would use the regression equation that we created to help us identify a predicted value of math achievement. This is what we discussed above and identified a predicted math achievement value of 60.438%. Since this is a predicted value, we would assume that the actual value of math achievement for that school might vary (either direction) from 60.438%.

The bold vertical line in the graph represents a range of possible values that may lie at some distance away (both below and above) from the regression line. Therefore we could think of a series of possible predicted math achievement outcome values that "spread out" around the predicted value we calculated (60.438%) using the regression equation. In this case the predicted value of 60.438 represents the "mean" of all the predictions possible when we use 40% as the value of the predictor.

To take one example from the scatter diagram and regression results shown above, we might take a look at one of the observations along this vertical line and see what its relationship might be to the predicted value of $Y$ (60.438%). As it happens, there is an actual case in the database of 40. You can see that it is the only sample case that falls on our vertical, prediction, line. This dot represents a school in the sample database with a F/R value of 40, and it roughly corresponds to a math achievement value of 43 (you can see this if you extend your eyes left to the $Y$ axis from the value of 40.

This happenstance allows us to see how our predicted value of math achievement (when FRLunch is 40 in the sample) relates to the actual sample value of math achievement in the database. For one thing, this points out the difference between predicted, population values, and sample values. We are using the regression equation to predict population values of math achievement. Even if we have the actual value of 40 represented in the sample, the population predicted value and the actual sample value might differ widely. This is due to the spread of the dots (cases) around the regression line. We are using the power of the correlation to influence the regression equation to help us predict values of the $Y$ variable that *are not known* when $X$ is at a certain value. However, since there is not a perfect correlation between math achievement and FRLunch, our prediction will represent a value that can be smaller or larger, depending on the extent of the scatter around the regression line.

If we add a couple of additional lines to our graph, we can observe how a correlation value affects these relationships. Figure 5.9 shows two additional lines on the scatter diagram in Figure 5.8. These lines "capture" the actual cases, or most of them, at this value of FRLunch. These lines represent two standard errors (of estimate) above and below our predicted value of $Y$ (60.438%). We created these lines using the "standard error of the estimate" of 12.64 reported on the SPSS® output in Table 5.5 (see the "Model Summary" panel). You may recall from introductory statistics courses that the standard error of the estimate is a measure of estimated variance of the population based on sample values. Thus two standard errors above and below the mean of the predicted value will represent the majority (near 95%) of the cases inside "normal" boundaries (of the sampling distribution).

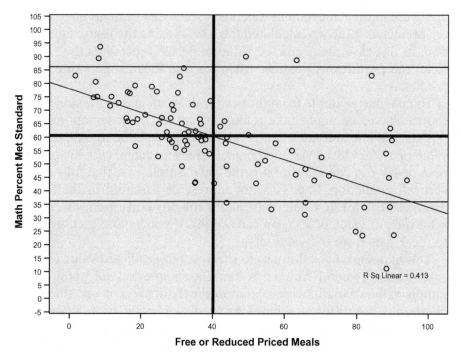

**Figure 5.9** Adding "capture lines" to the scatter diagram using the standard error of estimate.

If we add two standard errors of estimate to the predicted value, and subtract two standard errors from the prediction, we will obtain a range of values that will likely "capture" the true value of math achievement in the population when FRLunch is 40. Thus, in the majority of cases in the population, the actual value of math achievement will fall between 35.16% and 85.72%:

$$85.72\% = 60.438 + 2\,(\text{SEest}) \quad (60.438 + 25.28)$$
$$35.16\% = 60.438 - 2\,(\text{SEest}) \quad (60.438 - 25.28)$$

## The Confidence Interval

To be precise in our estimate of the predicted value, we would use the standard error of estimate to create the "confidence interval," which we discussed above as being the boundaries around the predicted value that would capture the true population value at a .95 probability (or whichever level of probability is chosen). In this case we would use the formula for confidence interval that we discussed earlier in this chapter:

$$CI_{.95} = z[SE(est)] + Y'$$

So we can substitute the values of SE(est) (12.64) and $Y'$ (60.438%) in the formula along with +/– $Z$ (+/– 1.96) as follows:

$$CI_{.95} = (+1.96)(12.64) + 60.438 = 85.2\%$$

$$CI_{.95} = (-1.96)(12.64) + 60.438 = 35.7\%$$

These results show that we can be approximately 95% confident that the "true" value of $Y$ lies between 35.7% and 85.2% in the population. You will note that these figures are not exactly equivalent to the figures we created above with just the standard error of estimate, since the confidence interval calculations represent +/–1.96 times the standard error of estimate rather than the two standard of errors we used as an example.

Now, if we return to the sample value of 40 in the database, we can make other discoveries about the data. From the analyses above we are confident that the predicted value of math achievement (60.438%) will be represented between population values of 85.2% and 35.7%. As it turns out, the sample value of 43 (approximately where the FRLunch sample value of 40 falls on the math achievement axis) is in this range of values. The fact that it is at the "low side of the scatter" on the diagram expresses this visually.

The size of the correlation affects the distance between the two "capture" lines on the graph. When the correlation between the two variables is stronger, the scores in the scatter diagram are closer to the regression line, with a resulting increase in accuracy of predicting values of $Y$.

By the way, the distance between the actual value of math achieve-ment and 60.438, the predicted value, is the "residual" value that we discussed earlier, and it is calculated for each predicted value of $Y$ by SPSS®. Most of the time you do not know the actual value of $Y$, since it may not be available in your data (which is the reason you use regression to predict!). In our case, as it just so happened, we had an actual $X$ value of 40 that corresponded to a $Y$ value of 43. Even though the 43 was a sample value, and may not represent the actual population value, it still points out how an actual value can differ from the predicted value. In this case, the regression line is sufficiently far away from our example case of an F/R value of 40 that it isn't terribly accurate in predicting the actual $Y$ value. Since the mean of $Y$ in this set of data is 59.77, the mean is slightly closer to the actual value of $Y$ (when F/R = 40) than is the predicted value using the regression analysis.

However, other values of F/R produce predicted $Y$ values that may be closer to the actual $Y$ values. For example, if we had used an F/R value of 21 from the graph (with a corresponding math achievement score of approximately 68), it would have fallen very near to the regression line, with a predicted value of 68.76%. Suppose that the actual value of math achievement was 68, and not just the value of the sample case; then the residual would have been very small.

## DETAIL—FOR THE CURIOUS

### Assumptions of Regression

As in the case of correlation, there are assumptions for linear regression that, when met, will likely yield the best estimates that reflect the (unobserved) population. Different textbooks look at these assumptions differently, and may include additional assumptions, so it is important to investigate several authorities according to the nature of your data. I will list some common assumptions here.

It is important to remember that if we are trying to make attributions about a population of values based on what we find from the sample, then we need to make sure that we have a randomly chosen sample. As I noted in the section on correlation, this will ensure that errors and individual differences are distributed equally and reduce bias.

1. The variables should be *normally distributed* (see the discussion of correlation above).
2. The variables should represent population variables that have *equal variances*. This is what researchers mean by "homoscedasticity."

You can check on homoscedasticity by looking at a scattergram. If the case values scatter around the line approximately equally, then the variances are probably equal. If, however, there are patterns other than this in the distribution of case values around the line, there may be violations of this assumption. For example, case values may be very close to the line at one end and far from the line at the other end (i.e., they might "bell out"), or the scatter diagram may represent a "cigar" pattern with more widely distributed case values in the middle of the line than at either end. In these situations, known as "heteroscedasticity," the standard errors would be affected, which would affect the confidence intervals and any significance tests.

If you examine the last scatter diagram (Figure 5.9), there appears to be a "bell-out" pattern to the case values such that they scatter more widely around the right end of the regression line than around the left end. This is a suspicious pattern and should be checked. (As we will see in a later section, however, this is probably due to a few outlying scores.)

3. The *standardized residuals should be normally distributed.* If you create a histogram of the standardized residuals (which can be saved from SPSS® analyses), the result should be a normal distribution, since the residuals are what remains after the relationship between predictor(s) and outcome variables are taken into account. There are other ways of detecting violations of this assumption, so you can check with detailed statistical treatments of regression assumptions if you suspect problems with your data.

## Fixed and Random Effects Modeling

Many statisticians make a distinction between "fixed-effects" and "random-effects" modeling, which can have an impact on the nature of the assumptions, strictly speaking. Fixed-effects modeling is where the predictor variable ($X$) has a set number of categories, like gender or experimental method, where the same categories are used in all situations. In contrast, in random effects modeling—as is most common in evaluation research and other nonexperimental analyses—the evaluator takes a random set of observations, with the categories or levels of the predictor variable ($X$) "emerging" from and determined by the data sampled. The categories and values of $x$ used in these situations are thus random.[15]

## Nonlinear Correlation

In the earlier section on correlation, I mentioned that some correlations could be based on nonlinear relationships. The "Curve Estimation"

---

[15] In most cases evaluation researchers do not follow the strict assumptions of fixed-effects modeling in non-experimental designs.

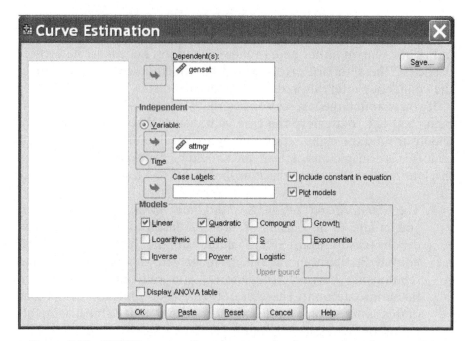

**Figure 5.10** SPSS® menu options for assessing linear and nonlinear models.

menu in SPSS® is a function that can help evaluators to estimate the nature of the relationship between two variables. If you choose the menu options "Analyze–Regression–Curve Estimation," you will get the screen shown in Figure 5.10, which will allow you to assess the bivariate linearity.

The SPSS® screen in Figure 5.10 shows several options that you can use to assess the nature of the relationship between our study variables. As you can see, you can place one outcome variable in the "Dependent" window and a predictor variable in the "Independent" window. Since the causal relationship could go either way, you will need to think through a reasonable causal assumption for deciding which variable is considered "dependent" or "independent." In our case we are assuming that job satisfaction is at least partially a function of the workers' attitudes toward management. Thus we have chosen as an outcome variable, "gensat," a measure of job satisfaction among a group of production workers. The predictor variable "attmgr" is a measure of the same workers' attitudes toward the management role in the manufacturing plant. (We will revisit this database in later sections.)

Under the "Models" panel in Figure 5.10, you can see that several modeling possibilities exist for you to examine your data. There are several possible patterns of relationships between two variables. We have make the assumption so far that the study variables are linear, that is, where the values of both variables lie on a straight line. However, there are other possible relationships where study values fall in different patterns. If you recall the example in Figure 4.2 between test scores and anxiety, you will see an example of a curvilinear relationship. In this case low to medium anxiety scores are related to low to medium test scores, but higher anxiety scores tend to "bend" the test scores back from medium to low in an inverted U shape.

The U shape may respond to a "quadratic" model, since the predicted $Y$ values would be created by combining two estimated coefficients: the coefficient for the values of the $X$ variable, and the coefficient for the squared values of the $X$ variable. The resulting equation would therefore represent a line that curves, or contains two different trajectories. The regression equation that we discussed above would change to include the added coefficient as follows:

Linear equation:

$$\hat{Y} = b_0 + b_1 X_1$$

Quadratic equation:

$$\hat{Y} = b_0 + b_1 X_1 + b_2 X_1^2$$

The SPSS® "Curve Estimation" panel indicates several possible relationships between the study variables. I have chosen "linear" and "quadratic"[16] for this example. We can first look at the data to be estimated by simply creating a scatter graph in SPSS®. The scatter diagram shown in Figure 5.11 shows the relationship between "gensat" and "attmgr" without displaying the regression line. As you can see, the data points follow a fairly stable pattern from upper left to bottom right. In the absence of other information, we might interpret this scatter pattern as a strong inverse correlation with some "outliers" on the right side of the graph that could distort the linear regression line.

---

[16] The "quadratic" model indicates that we want to add the squared values of "attmgr" to the regression equation along with the unsquared values of "attmgr" in predicting "gensat." This way we can see if the squared values add information to estimating the curve.

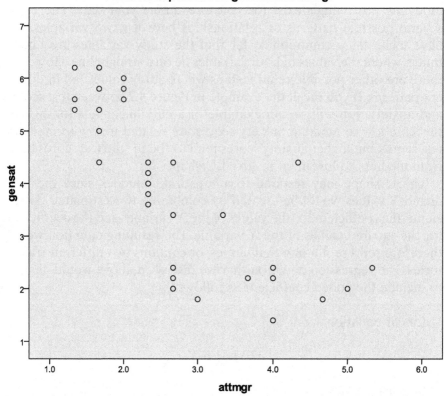

**Figure 5.11**    Scatter diagram between "gensat" and "attmgr" without regression line.

However, other possibilities exist. One such possibility is that there is a different (nonlinear) pattern that fits the data better. Figure 5.12 shows the same scatter diagram, but this time with the two models specified earlier (i.e., linear and quadratic) in the Curve Estimation procedure in SPSS®. The SPSS® output in Table 5.7 indicates that the quadratic model may be a better fit to the data.

The output in Table 5.7 shows that both models are statistically significant (see the "Sig." column under "Model Summary"). However, the $R^2$ for the quadratic model accounts for about 10% more of the variance in job satisfaction ("gensat") than the linear model ($.737 - .636 = .101$) among this sample of workers. The resulting regression equations would be as follows:

Linear equation:

$$\hat{Y} = 6.822 - 1.087 X_1$$

**Linear and Quadratic Relationships between gensat and attmgr**

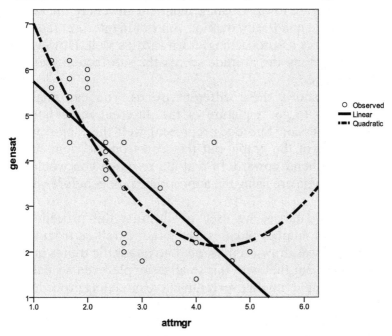

**Figure 5.12**   Scatter diagram including the regression line.

**Table 5.7   Linear and Quadratic Estimates for Model _F_ Fit Assessment**

Model Summary and Parameter Estimates

Dependent Variable: gensat

| Equation | Model Summary | | | | | Parameter Estimates | | |
|---|---|---|---|---|---|---|---|---|
| | R Square | F | df1 | df2 | Sig. | Constant | b1 | b2 |
| Linear | .636 | 64.702 | 1 | 37 | .000 | 6.822 | −1.087 | |
| Quadratic | .737 | 50.463 | 2 | 36 | .000 | 10.233 | −3.629 | .406 |

The independent variable is attmgr.

Quadratic equation:

$$\hat{Y} = 10.233 - 3.629X_1 + .406X_1^2$$

You can see from the equations that the interpretation of the data would be quite different. A linear relationship (based on the linear equation) would suggest a steady upper left to lower right pattern to the data along a straight line. Thus job satisfaction would have a direct, inverse, relationship with attitudes toward management (i.e., the stronger the attitudes are toward management, the lower are the satisfaction scores). From the quadratic equation you would interpret the predicted satisfaction scores as proceeding downward as well. However, past a certain point among the attitude scores, the satisfaction scores would begin to increase.

Figure 5.12 shows these different trends. You can compare the regression lines to get a picture of the different regression models. The linear regression line does represent well the majority of cases on the left side of the graph, but the scores on the right side of the graph begin to bend upward. In real-life research you would want to make sure that you are using the appropriate model before you finalize your analyses.

This sample database we used to illustrate this procedure represented a limited number of cases. The larger database from which our limited sample was drawn represented no quadratic trends in the data. We might point out that with this small a sample, even a small number of "deviant" samples might overly affect the modeling procedure. If you look at Figure 5.12 again, you might see that one case (attmgr approximately equal to 4.3, and satisfaction about 4.5) may strongly influence the different model estimates. In actual research studies evaluators would want to consider such cases carefully before drawing conclusions about the linearity of the resulting data models. This influence of "outlying" cases is the subject of chapter 6 where we devote considerable attention to making sure that our data meet the research assumptions before we conduct an evaluation analysis.

## Calculating the Standard Error of the Estimate

I mentioned earlier that the value of the SE(est) can range from 0 to the value of $SD_Y$. It is helpful to remember this when you calculate SE(est); if you get a negative value, or a value that exceeds the value of $SD_y$, then you have made a mistake. The SPSS® output will identify this value, but it is sometimes helpful to look for it and do a quick calculation to see the relationship between $R^2$ and the SE(est). The example in Table 5.8 uses the formula I introduced earlier with findings from a worker participation database that will be discussed in a later chapter. For now, the outcome variable "gensat" is a measure of general job satisfaction among a group of workers, and the predictor "attmgr" is a measure of attitudes toward management by those same workers. So $SD_y = 1.278$, and as you can see from the SPSS® output in Table 5.8, the $r = .35$.

**Table 5.8    SPSS® Output for Example Data of Job Satisfaction and Attitudes Toward Management**

| | | | | Model Summary | | | | | |
|---|---|---|---|---|---|---|---|---|---|
| | | | | | | Change Statistics | | | |
| Model | R | R Square | Adjusted R Square | Std. Error of the Estimate | R Square Change | F Change | df1 | df2 | Sig. F Change |
| 1 | .350ᵃ | .123 | .115 | 1.20231 | .123 | 16.477 | 1 | 118 | .000 |

a. Predictors: (Constant), attmgr.

Using the following formula, we can calculate the standard error of estimate, which in the output in Table 5.8 is listed as 1.20231 (see the fourth column of data).

$$SE(est) = SD_Y \sqrt{1 - r^2}$$

If you use this formula, you will verify that the SE(est) is 1.20 as shown in output of Table 5.8. The only thing to remember is that you need to use the *adjusted* $R^2$ in the calculation rather than the actual $R^2$, since sample size adjustments affect the overall $R^2$ calculation.

## DISCOVERY LEARNING

Let us review what we have found thus far with respect to a simple regression using F/R to predict math achievement levels. In this example we saw that the correlation between the two variables enabled us to predict one variable from another. As we saw, a stronger correlation produced a better or more accurate prediction. Once again, we have discovered that using visual representations of the data and the analyses is helpful for understanding the nature of the patterns embedded in the data. In particular, we observed that the "line of best fit" is a visual referent for the scatter of the data, the formula for which can describe the slope and intercept of the line. The regression coefficient $\beta$ (or "beta") describes the extent to which the value of the outcome variable changes for a given unit change in the value of the predictor variable.

We confirmed substantively the pattern we discovered earlier, that school level math achievement is inversely related to the percentage of students at the school who qualify for free or reduced price meals.

Using this relationship in a regression equation, we can predict values of a school's math achievement level when we know values of F/R:

- Correlation enables prediction through regression.
- Correlation helps to account for variance in the outcome variable.
- "Residuals" are helpful for illustrating the interpretation of the variance components of the regression equation.

## Terms and Concepts

**Adjusted $R^2$**  A value showing the likelihood of the $R^2$ value that would be obtained with all the data from the actual population in the analysis. A correlation or $R$ value based on a limited sample size, with only one predictor, may result in a larger correlation than if all the values of the population were involved. The adjustment to $R$ takes into account sample size and number of predictors to provide a value that would more likely reflect a population-based analysis.

**ANOVA table in regression results**  The table that analyzes the relationship among the components of regression in order to assess the model "fit" of the predictor and outcome variables. This analysis produces an "$F$ test" that expresses how large the regression component is relative to the error component of a sum of squares analysis. The larger the $F$ value, the larger the regression component is relative to the error component, and therefore the greater the fit of the model. ANOVA (analysis of variance) results express the likelihood of differences among a series of sample group means.

**Beta**  This measures is the slope of the regression line in a regression analysis. When standardized, it is expressed in standard normal ($Z$) values, and represents the change in the outcome variable associated with a one standard deviation change in the predictor variable, when other predictor variables in the analysis are controlled. *Standardized* beta is typically expressed by the Greek $\beta$ and can be called "beta weight" or "regression weight." *Unstandardized* beta is also the slope of the regression line, but its value retains the metric of the variables in the analysis, without transformation to a standard normal scale. In this form the beta is typically referred to simply as a regression coefficient.

**Confidence intervals**  The boundaries within which the true population values are likely to fall, given the information derived from sample values.

**Curve estimation**    The SPSS® procedure that allows an evaluator to assess whether the data being analyzed are linear, or better approximated by several nonlinear estimation procedures (quadratic, exponential, etc.).

**Homoscedasticity**    The assumption for regression that ensures comparable variances in the outcome variable for similar values of the predictor, and therefore "balanced" patterns of scatter among the cases around the regression line. Heterscedasticity is the violation of this assumption.

**Line of best fit**    The line in a scatter diagram representing the relationship among an outcome and predictors calculated such that the sum of the squared distances between each point and the line is as small as possible.

**Omnibus test**    The overall analysis in regression that expresses the extent to which the regression model as a whole is significant. This is represented in the ANOVA summary analysis of an SPSS® output file.

**$R^2$ change statistic**    Provides information about how the overall $R^2$ for the model changes as additional independent variables are added to the analysis.

**Regression**    A statistical analysis that allows the evaluator to predict values on an outcome variable given knowledge of the values of a predictor, or multiple predictor, variables. Bivariate regression uses a single predictor, whereas "multiple regression" implies more than one predictor. Regression techniques can focus on linear relationships between outcome and predictors where the best-fitting line is a straight line (multiple linear regression); nonlinear regression assumes relationships between outcome and predictors not based on a straight best-fitting line.

**Regression formula**    The equation of the regression line that expresses the values of the slope, intercept, and error, taking into account the predictors and outcome of the analysis.

**Residual scores**    These scores represent the difference between predicted scores and actual scores. (Residual scores are often referred to as "error" when compared to the "regression" values, which are the differences between the predicted scores and the mean of the outcome variable.)

**Standard error of estimate**    The population estimate of the variation of the data points around the regression line. Stated differently, it represents the distribution of the differences between our *estimated* values and the *actual* values of the data points.

**Sum of squares**   The accumulated squared deviations of scores from their mean. This is a rough index of variation in that the larger the number, the greater is the difference of the scores from their means. The variance is obtained by dividing this measure by $N$ or $N - 1$, depending on the nature of the analysis.

*t* **Test in SPSS® regression results**   This value indicates the probability that the slope of the regression line resulting from the relationship between the outcome and predictor is 0 or flat.

### Real World Lab—Bivariate Regression

Use SPSS® to conduct a bivariate regression with the data presented in our discussion of correlation in Chapter 4. Regress the outcome variable (Reading Achievement) on the predictor variable (STAR_mean):

1. What is the overall $R^2$?
2. Is the "omnibus" finding significant?
3. Is the test of the predictor variable significant?
4. Use the sum of squares in the ANOVA results to create $R^2$.
5. What is the regression formula?
6. Provide a scatter diagram of the results.
7. Express the regression results in the language of the research problem.
8. Is the relationship linear?

The answers to these practical application questions are in Chapter 12, the last chapter of the book.

# 6

# CLEANING THE DATA— DETECTING OUTLIERS

Discovering patterns of relationships among data requires an accurate picture of the ways in which data are actually "arrayed." When we examined the scatter diagrams earlier, we were able to see how the data scattered out around a line of best fit. We observed that the stronger the correlation, the tighter is the array of the scores around the line. We also observed that removing just a few data points can meaningfully affect the overall relationship.

The results of a statistical analysis are affected by all the cases in the database. If some of those cases are accidentally lost, or if some are extremely different from the majority of other cases, we may observe quite different results. It is important that we examine each of the cases carefully to make sure that they all "belong" to the set of data. Often, we may observe that there are some data points that are scattered far away from the other cluster of data points. These *outliers* are important to examine because they may exert a good deal of influence on the outcome of the analysis.

There are many questions to consider about outliers. They may accurately represent a given data point. That is, they may represent the very few possibilities that show up when you deal with normally distributed data; some observations will fall in the outer tails of the distribution. In this case the researcher must decide what to do with them. If they

*The Program Evaluation Prism: Using Statistical Methods to Discover Patterns,*
by Martin Lee Abbott
Copyright © 2010 John Wiley & Sons, Inc.

are "real" data points and not the result of faulty data entry or are responses far removed from the expected range, the researcher needs to think about what the impact might be of deleting them, or keeping them in the analysis. If there are many such data points, the decision is more difficult, but even a few pose serious questions. The researcher has several options:

- Should I delete the data point(s) on the ground that it represents an extreme score that unduly affects the overall relationship among the study variables? Does it detract or distort the overall relationship?
- Should I replace extreme scores with "less extreme" scores or mean scores (or some other replacement scheme) so that I can help maintain the relationship without losing all of the information from the data point?
- Can I "transform" the scores in some fashion so that the resulting array of scores is more within a normally distributed range?

Outliers may not actually affect the relationship of two variables even though they represent extreme scores. As one example, if they lie on the same "path" as the cluster of scores, but just further away from the pack of other scores, they may not substantially change the outcome of the analysis. Or, if the extreme scores are lateral to the cluster of dots but not so much that they impact the outcome, then their disruption may be contained.

These and other considerations are addressed thoroughly by several comprehensive treatments of multivariate statistics, such as that by Tabachnick and Fidell (2007), Pedhazur (1997), and, Judd and McClelland (1989). We can only begin to address the primary issues that affect extreme scores in this book. If you are interested in pursuing the intricacies of outliers and how to detect and remedy problematic data, you should consult these sources. I would only add here that serious researchers should at least perform cursory examinations of a database before they engage in their analyses. We will discuss some of these processes next.

## UNIVARIATE EXTREME SCORES

Evaluators can profit from a statistical program like SPSS® to deal with extreme scores in a database. It is important to examine both *univariate extreme scores* (one variable) and/or *multivariate extreme scores* (more than one variable in an analysis such as regression). Univariate outliers can be detected by visual means through some of the SPSS® graphics programs, but can just as easily be seen by examining common SPSS® output that you would get from running a descriptive analysis on variables of interest.

You should first check the standard output to see whether or not a variable you might use is *skewed* or otherwise not normally distributed. Histograms are good for this purpose, but consult the *skewness* coefficients to see whether, or how much, the distribution of the variable may depart from normality.[17] Correlation is considered a "robust" procedure, or one that can tolerate some violation of assumptions. Therefore, if a variable is slightly skewed, it may not be problematic to use it in a correlation procedure.

Another way to observe extreme scores for individual variables is to use standard scores ($Z$ scores) for each of the values. SPSS® will create these if you simply check the box that requests saving standard scores when you use the "Descriptives" command. You can examine the standard scores (at both positive and negative ends) and look at scores of +/− 3.00 as extreme scores. Different researchers use different $Z$ "cutoff" scores, but the important point to remember is that you are simply trying to examine the scores for any that may be problematic to the analysis. Using a cutoff value in a mechanical fashion precludes the possibility of a thorough consideration of why the score is what it is and how it might affect the overall analysis. If you identify extreme values using this procedure, you will need to think about the issues we discussed above: How many extreme scores are there? Are these "real" scores? How would it impact the analysis to delete or change them?

To take an example using the sample data described above, one math achievement score (of about 11%) showed a standard score of −2.95, which is just below the cutoff we suggested earlier. This value on the scatter diagram is highlighted in the right lower corner in Figure 6.1. Identifying this score in this manner does not suggest that the score should be deleted or changed. Rather, you need to explore the database to determine whether it is an actual score, or if there are reasons to suspect it should be dropped from the analysis. In this instance it is an extreme case, but it appears to lie in the same direction of the "pack" of other scores. In the next section we will discuss additional methods to determine how it might affect the overall analysis.

---

[17] You can determine whether a variable is skewed by performing a test on the figures provided by SPSS® wherein you divide the skewness coefficient by its standard error. If the result is larger than the typical .05 level of significance for $z$ scores, +/−1.96, the variable can be considered skewed. Other statisticians, such as the sources listed above, use different cutoff values that can be considered as well.

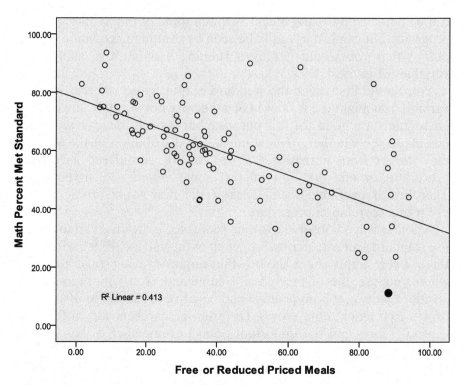

**Figure 6.1** Scatter diagram identifying the potential impact of extreme scores.

## MULTIVARIATE EXTREME SCORES

The next step is to look for extreme scores when we analyze variables *together* rather than looking at them individually. Checking for multivariate outliers is a bit more complex, but the methods can help to assess the extent to which extreme scores might impact the results of procedures like regression. These procedures are available for regression problems with more than one independent variable, but using an example from a simple regression procedure will illustrate how they work.

As you may recall, we discussed residual scores above as a way of understanding how scores disperse around the regression line. Residual scores can be an excellent way of looking for extreme scores because large residual values indicate large distances between what we predict from a regression equation and the actual value used to create it. Obviously this has a great deal to do with the size of the correlation between the variables, but we can examine the residuals to see where

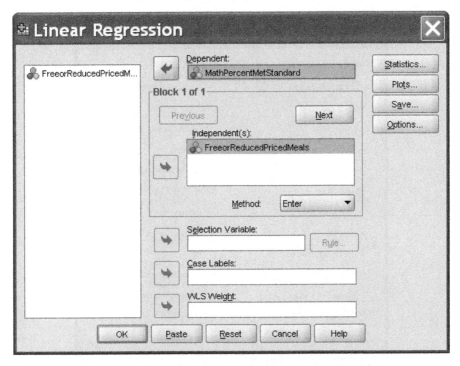

**Figure 6.2** SPSS® menu for specifying linear regression.

the extreme values are and whether these will likely be problematic to our results.

While the *unstandardized residuals* are helpful, it is important to remember that each may have different variances. Thus SPSS® provides a list of *standardized residuals* where each residual (the difference between the actual and predicted y values) has been divided by the standard deviation of the residuals (which is also the standard error of estimate).

The way to obtain the scores we will discuss in this section is through the SPSS® specification for running a regression analysis. If you choose "Analyze–Regression–Linear" you will get the SPSS® field shown in Figure 6.2. As you can see, we are specifying math achievement as the outcome variable and F/R as the predictor. On the right side row of buttons on this panel, you will find "Save," which allows you to select scores from the analysis you are running and save them in your database.

If you select "Save," you will see the screen in Figure 6.3. As shown, I have selected several options that will illustrate some of the measures we will use in this section. By selecting "continue," you can complete

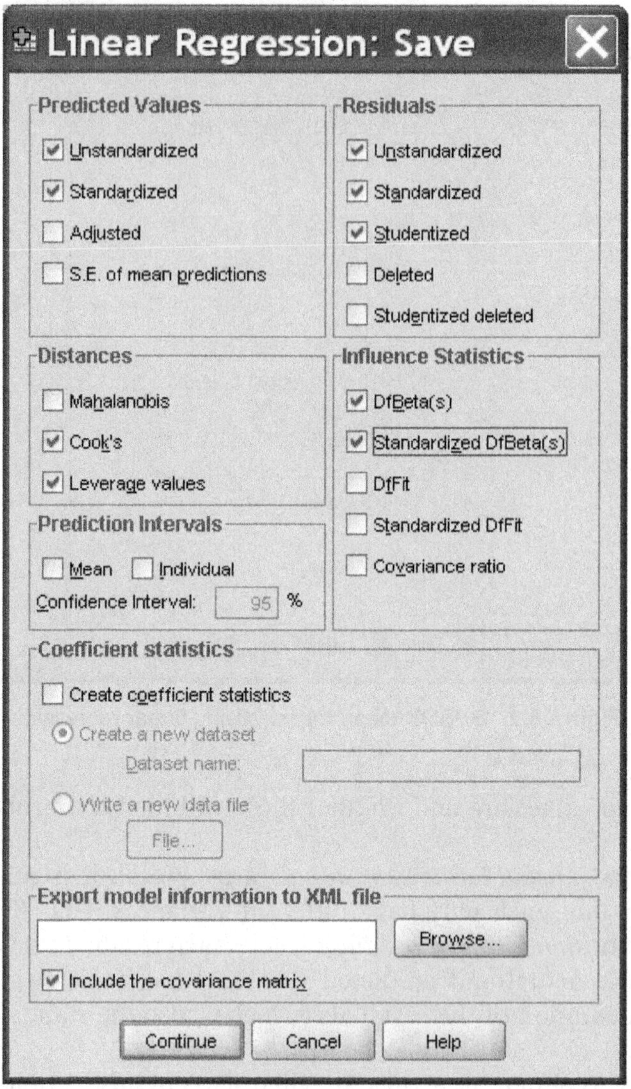

**Figure 6.3**   SPSS® menu for saving values that assess extreme scores.

the analysis. The output file will list the regression outcomes, but more important, the values you specified will be added as separate variables in the data file. Figure 6.4 shows the resulting data file with the added variables.

Figure 6.4 shows the data file that results from running the analysis of the "Linear Regression–Save" choices that we specified above. In addition to the output file showing the results of the statistical analysis, SPSS® will save a series of variables in the data file according to

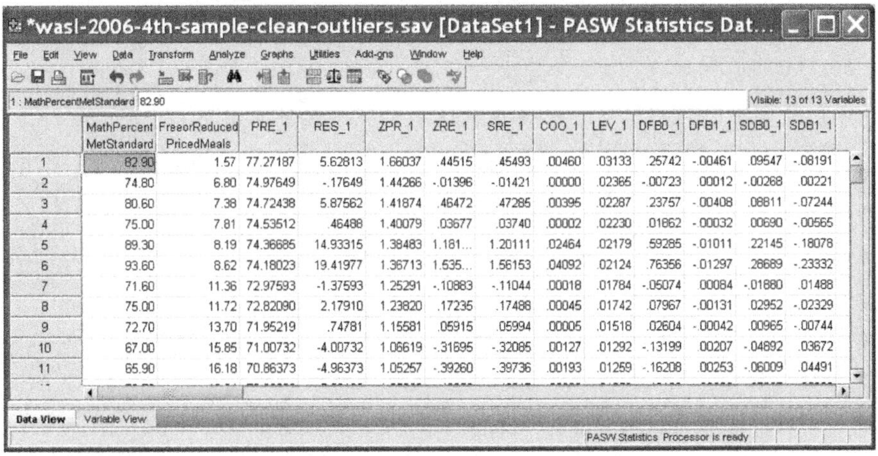

| | MathPercent MetStandard | FreeorReduced PricedMeals | PRE_1 | RES_1 | ZPR_1 | ZRE_1 | SRE_1 | COO_1 | LEV_1 | DFB0_1 | DFB1_1 | SDB0_1 | SDB1_1 |
|---|---|---|---|---|---|---|---|---|---|---|---|---|---|
| 1 | 82.90 | 1.57 | 77.27187 | 5.62813 | 1.66037 | .44515 | .45493 | .00460 | .03133 | .25742 | -.00461 | .09547 | -.08191 |
| 2 | 74.80 | 6.80 | 74.97649 | -.17649 | 1.44266 | -.01396 | -.01421 | .00000 | .02365 | -.00723 | .00012 | -.00268 | .00221 |
| 3 | 80.60 | 7.38 | 74.72438 | 5.87562 | 1.41874 | .46472 | .47285 | .00395 | .02287 | .23757 | -.00408 | .08811 | -.07244 |
| 4 | 75.00 | 7.81 | 74.53512 | .46488 | 1.40079 | .03677 | .03740 | .00002 | .02230 | .01862 | -.00032 | .00690 | -.00565 |
| 5 | 89.30 | 8.19 | 74.36685 | 14.93315 | 1.38483 | 1.181... | 1.20111 | .02464 | .02179 | .59285 | -.01011 | .22145 | -.18078 |
| 6 | 93.60 | 8.62 | 74.18023 | 19.41977 | 1.36713 | 1.535... | 1.56153 | .04092 | .02124 | .76356 | -.01297 | .28689 | -.23332 |
| 7 | 71.60 | 11.36 | 72.97593 | -1.37593 | 1.25291 | -.10883 | -.11044 | .00018 | .01784 | -.05074 | .00084 | -.01880 | .01488 |
| 8 | 75.00 | 11.72 | 72.82090 | 2.17910 | 1.23820 | .17235 | .17488 | .00045 | .01742 | .07967 | -.00131 | .02952 | -.02329 |
| 9 | 72.70 | 13.70 | 71.95219 | .74781 | 1.15581 | .05915 | .05994 | .00005 | .01518 | .02604 | -.00042 | .00965 | -.00744 |
| 10 | 67.00 | 15.85 | 71.00732 | -4.00732 | 1.06619 | -.31695 | -.32085 | .00127 | .01292 | -.13199 | .00207 | -.04892 | .03672 |
| 11 | 65.90 | 16.18 | 70.86373 | -4.96373 | 1.05257 | -.39260 | -.39736 | .00193 | .01259 | -.16208 | .00253 | -.06009 | .04491 |

**Figure 6.4**    Revised data file including outlier diagnostic variables.

what you have requested. We will refer to this output as we discuss the values we have chosen.

The first four variables created by our analysis are the following that you can identify in Figure 6.4:

"PRE_1" is the unstandardized predicted values ($Y'$) for each case.

"RES_1" is the unstandardized residuals for each of the cases.

"ZPR_1" is the standardized predicted values $Z_{Y'}$ for each case.

"ZRE_1" is the standardized residuals for each case.

When I sorted the values on the standardized residuals ("ZRE_1") variable (using "Data–Sort Cases"), I found three cases that were at or near the cutoff value we described above for standard scores (+/–3.00). Since we are dealing with standardized values (residual values), we can use this cutoff criterion. Figure 6.5 shows these three values.

Two of the values were beyond +/– 3 standard scores, and the third was included for demonstration, even though it was below the cutoff criterion (ZRE_1 = 2.66). Obviously other cutoff scores could be identified that would locate other extreme scores, so it is important to set a cutoff value according to your own notions about what should constitute "extreme" for a given research project, and to use your judgment about what scores to examine. As shown by the highlighted dots in the Figure 6.5, these values are each well "above the pack" of the main cluster of scores and may thus inordinately affect the slope of the regression line.

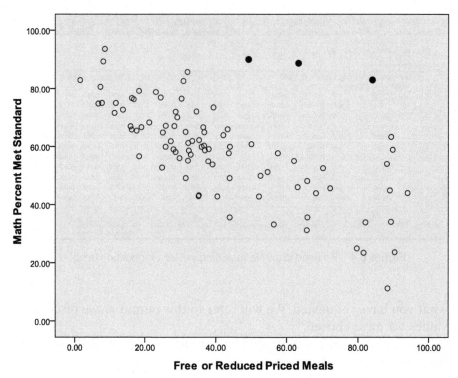

**Figure 6.5**   Scatter diagram identifying extreme scores by standardized residuals.

SPSS® also provides *Studentized* residuals ("as SRE_1"). These measures result from dividing each residual by its estimated standard deviation rather than the standard error of estimate to account for the fact that individual residuals may have unequal variances. This measure also identified the same three scores in our example, as shown in Figure 6.5, so we can examine the scores to see how extreme they are and whether we want to retain them, delete them, or somehow change them. Visually it appears that they are somewhat "lateral" to the pack and might represent an impact on the regression equation.

## DISTANCE STATISTICS

In trying to decide how extreme scores may be, SPSS® provides several "distance" measures. That is, several measures are calculated for each score that indicate how scores lie in relation to the other scores in the regression scatter diagram. These "distances" measures are available on the *Linear Regression: Save* and will be saved to the data file as follows:

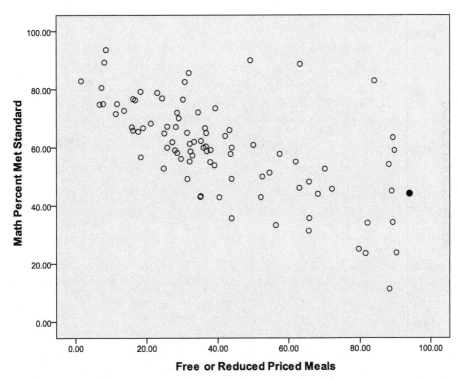

**Figure 6.6**  Scatter diagram identifying extreme scores using centered leverage values.

"LEV_1" are centered leverage values.

"COO_1" are Cook's distance values.

*Centered leverage values* ("Leverage values" in Figure 6.3) indicate these distances for the independent variable. Other measures describe extreme distances for both independent and dependent variables. The greater the distance between each $X$ value and the mean of $X$, the larger is the centered leverage value, whereas values closer to the mean of $X$ are increasingly smaller. Statisticians differ in their opinions of cutoff criteria based on these values. One way of examining this measure is to look at centered leverage values that represent 3 standard deviations away from the centered leverage mean.

Using this criterion, I found that one value in the database was considered extreme ($X = 94$, centered leverage value = .054).[18] As shown in Figure 6.6, it has the greatest "distance," but is generally

---

[18]With these data, the mean "centered leverage value" was .01 with an SD of .014, resulting in a cutoff value of .053.

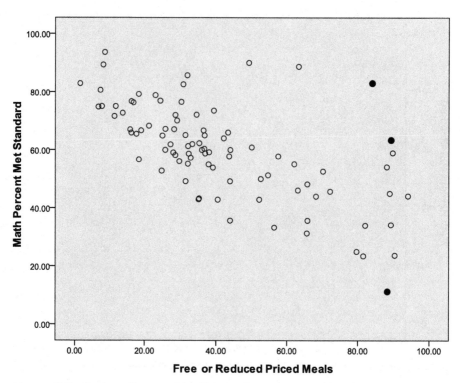

**Figure 6.7** Scatter diagram identifying extreme scores using Cook's distance values.

in line with the pattern of the pack. Less stringent criteria may identify other potentially problematic $X$ values that might be more disruptive.

*Cook's distance values* ("COO_1") measure the effect on all the residuals if a particular case were dropped from the analysis. It therefore is a measure of both independent and dependent variables, with large values indicating more impact on the regression coefficients. Here again, cutoff criteria are debated by various researchers. One way to examine Cook's distance values visually is to sort the database by the values and to identify how extreme they become. They range from small values of 0 to larger values where the extreme scores may be several times larger than the immediately preceding scores.

Using this method, we can select several scores that might appear to be quite distant from the others. In this database I identified three scores shown in Figure 6.7. This visual identification of the extreme values may be helpful to see how the scores may impact the regression equation, but we would need to use additional measures to help us more specifically make these assessments. We will discuss these in the next section. Just using the graphed values, however, we might explore the possibility of temporarily deleting these scores to see what the impact might be.

## INFLUENCE STATISTICS

As we have noted, it is important to examine the data for extreme scores, since they may have a large impact on the regression equation. Examining the residuals and using the distance statistics are helpful in locating potential extreme scores, but SPSS® provides other measures that identify the impact on regression coefficients *if individual scores are deleted from the analysis*. These measures are available in the "Linear Regression: Save" menu as "Influence Statistics" (see Figure 6.3).

*"DfBeta"* measures ("Dfbeta(s)" in the SPSS® menu) provide values that represent the change in the regression coefficients when a particular score is deleted from the analysis. If you select this box, SPSS® will save two scores: *"DFB0"* measures show what impact deleting a score would have on the *intercept* in the equation, whereas *"DFB1"* show the impact of deleting a score on the *slope* or regression coefficient. These measures essentially represent the amount of change in the intercept and slope of the regression line if you were to run the regression, delete the scores, and then re-run the regression line. If you subtract the values from the original regression, you will observe the change is equal to the DfBeta measures.

To take one example from the data we have examined, Figure 6.8 shows one score that has been identified as extreme through the use of standardized residuals and Cook's distance measures. Below is an examination of how this observation affects the regression equation if it were deleted. The overall (unstandardized) regression equation, based on the regression analysis, is

$$\text{MathAchievement} = (-.438)\, \text{F/R} + 77.958$$

The DFB0 and DFB1 values for the extreme score are $(-1.027)$ and $(.0366)$, respectively. If we were to delete this extreme score, the resulting regression equation changes to

$$\text{MathAchievement} = (-.475)\, \text{F/R} + 78.985$$

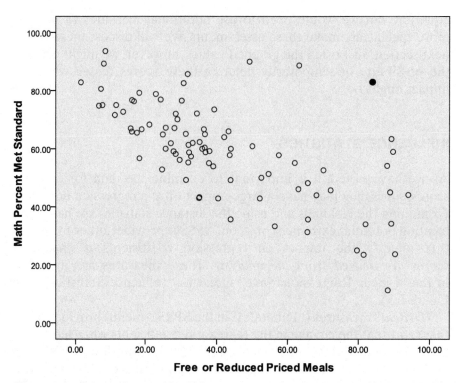

**Figure 6.8** Scatter diagram identifying an extreme score using standardized residuals and Cook's distance.

Thus, subtracting the DFB0 score (–1.027) from the initial intercept yields the value of the intercept shown in the revised regression equation (78.985):

$$(77.958) - (-.1.027) = (78.985).$$

Subtracting the DFB1 score (.0366) from the initial slope value (–.438) yields a value of (–.475), which is represented in the second regression equation where the extreme value has been deleted:

$$(-.438) - (.0366) = (-.475)$$

When we deleted this one extreme score, the (adjusted) $R^2$ value changed from .41 to .48 as shown in the two output tables of Table 6.1. The first table includes the extreme score and the second does not include the extreme score.

**Table 6.1  Model Summaries of Results with and without Deleted Score**

| Model Summary[b] | | | | |
|---|---|---|---|---|
| Model | R | R Square | Adjusted R Square | Std. Error of the Estimate |
| 1 | .643[a] | .413 | .406 | 12.64335 |

a. Predictors: (Constant), FreeorReducedPricedMeals
b. Dependent Variable: MathPercentMetStandard

| Model Summary[b] | | | | |
|---|---|---|---|---|
| Model | R | R Square | Adjusted R Square | Std. Error of the Estimate |
| 1 | .692[a] | .478 | .472 | 11.84691 |

a. Predictors: (Constant), FreeorReducedPricedMeals
b. Dependent Variable: MathPercentMetStandard

By deleting one extreme score, we explained an additional 7% of variance in math achievement! I mention this because it underscores a couple of things we have discussed thus far. First, extreme scores may make a huge difference to the analysis depending on where they are located. Second, it is important to use all the measures we have discussed for locating extreme scores as a way of identifying all the influences on the regression coefficients. We should not be overly concerned about technical cutoff criteria but rather with using all the outlier results to identify potential troublesome scores. Then we can assess whether or not they should be deleted, changed, or allowed to remain in the database.

*"Standardized DfBeta"* measures ("Standardized DfBeta(s)" in the SPSS® box) do the same thing as their unstandardized counterparts, except that they represent measures that have been standardized so that they take into account different scales of measurement in the independent and dependent variables. Selecting this option will save two additional scores to the database:

*"SDB1_1"* (standardized DfBeta for the slope)
"SDB0_1" (standardized DfBeta for the intercept)

These standardized measures are helpful in determining how large an impact it would represent on the regression coefficients if an extreme score were dropped from the analysis. SPSS® guidelines suggest that absolute SDB1 (or SDB0) scores of 2 divided by the square root of the number of cases will identify extreme scores. Other criteria have been suggested depending on the nature and size of the database.

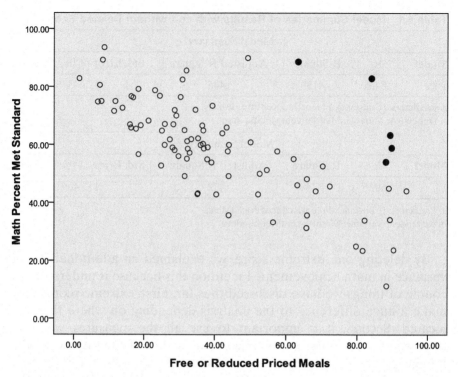

**Figure 6.9** Scatter diagram identifying extreme scores using standardized DfBeta values.

In particular, Cohen et al. (2003) suggest an adjusted cutoff criterion for smaller databases, and that cases with large values relative to other case values may have higher influence.

According to these considerations, I highlighted scores in Figure 6.9 that appeared to be high in comparison to other values when sorting the data by the SDB1 and SDB0 values. If deleted, the $R^2$ would increase from .41 to .62, a large change. Again, it is up to the evaluator to decide how to manage this information. Are these legitimate scores? Is the impact to the analysis large enough to warrant deleting them from the analysis?

## DISCOVERY LEARNING

Depending on the database, deleting scores for whatever reason can change the results to a great degree. Data are hardly ever "clean" in the sense that there are no anomalies present. That is the nature of actual data derived from evaluation studies. But to have a clearer sense of what the true relationships are among a set of study variables, we should attempt to identify and possibly "neutralize" the distortion caused by any extreme scores.

## Terms and Concepts

**Distance statistics**   SPSS® measures calculated for each score that indicate how scores lie in relation to the other scores in the regression scatter diagram:

- *"Centered leverage" values* SPSS® distance measures for the independent variable.
- *Cooks distance values* SPSS® distance measure of the effect on all the residuals if a particular case were dropped from the analysis. It therefore is a measure of both independent and dependent variables, with large values indicating more impact on the regression coefficients.

**Influence statistics**   SPSS® provides measures that identify the impact on regression coefficients if individual scores are deleted from the analysis.

- *"DfBeta" measures* SPSS® values that represent the change in the regression coefficients when a particular score is deleted from the analysis. Separate measures are provided for impact on the slope ("DFB1") and intercept ("DFB0").
- *"Standardized DfBeta" measures* Do the same thing as their unstandardized counterparts, except that they represent measures that have been standardized so that they take into account different scales of measurement in the independent and dependent variables.

**Outlier**   Data points that are scattered far away from the other cluster of data points in a graph or database. (Also referred to as "extreme scores.") Outlier scores can be *univariate* (belonging to the values of one variable) or *multivariate* depending on the nature of the analysis. Univariate outliers are those that stand apart among the other values of a single variable, whereas multivariate outliers stand apart from the cluster of other scores that contain the values of more than one variable, like a score among a scatter diagram array in a multiple regression analysis.

**Skewness**   Measure of a variable that expresses the extent to which it represents a balanced distribution. If it is a skewed distribution, more of the scores lie on one side of the distribution, creating a lopsided distribution. One or more extreme cases of the variable may be responsible for "skewing" the distribution.

**Standardized residual scores**   Residual scores in which each residual has been divided by the standard deviation of the residuals (which is also the standard error of estimate).

**Studentized residuals**   Residual values that result from dividing each residual by its estimated standard deviation rather than the standard error of estimate so as to account for the fact that individual residuals may have unequal variances.

**Unstandardized residual scores**   Difference between actual (observed) values and predicted values in a regression analysis using the metrics of the raw data variables.

### Real World Lab—Extreme Scores

Use the STAR Protocol database that you used for the last two practical application exercises. In this exercise, you will attempt to locate outlier scores in the database and discuss the results of your analysis.

Use the SPSS® command to "Analyze–DescriptiveStatistics–Descriptives" to save standardized (Z-score) values to identify *univariate outliers*:

1. Are any of the Z-score values beyond +/– 3.00?
2. Visually inspect a scatter diagram to identify any extreme scores and determine whether or not they are "close to the pack."
3. What is your plan with respect to any identified outliers?
    Conduct a regression analysis predicting reading achievement from STAR_mean scores and save the following as variables added to your database to help identify *multivariate outliers*: standardized and unstandardized residuals, DfBeta(s), and standardized DfBeta(s), Cook's distance values, and Leverage values.
4. What multivariate outliers can you identify?
5. What, if any, impact is there for deleting problematic outliers?
6. What is your overall plan to deal with multivariate outliers?

# 7

# MULTIPLE CORRELATION

## INTRODUCTION

Now that we have explored the uses of regression with one predictor variable, we return to the general topic of regression, but this time with more than one predictor. Multiple regression is common in the literature and very helpful to evaluators. In order to see its value, we will make reference to examples we used in the preceding chapters. The first of these was the situation where we were trying to use "time spent reading," by which we will explain the "reading achievement test scores" with the use of Venn diagrams.

Before turning to a detailed treatment of multiple regression, however, we will examine a couple of special cases of correlation that will help us understand better how adding variables to the regression model affects the outcome variance. Partial correlation and part correlation are very helpful processes that help illuminate the multiple relationships in a regression analysis.

We know that no one variable alone will explain all the variance of another; life is not that simple. Therefore we can try to understand how looking at several predictor variables analyzed at the same time might help account for more of the variance of an outcome variable. This is the essence of multiple regression. We are looking at how all the

*The Program Evaluation Prism: Using Statistical Methods to Discover Patterns,*
by Martin Lee Abbott
Copyright © 2010 John Wiley & Sons, Inc.

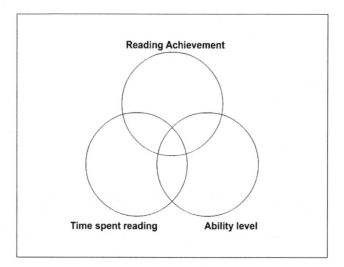

**Figure 7.1**   Adding predictor variables to account for variance in an outcome variable.

relationships together help account for the variance of the target variable. Thus, we might add "ability level" to our original regression analyses and see what additional explanation it provides, as shown in Figure 7.1.

We will never be able to explain all the variation in reading achievement in this manner because there are probably an infinite number of potential explanations. But adding additional variables helps us get closer to understanding why the original scores vary so much. With multiple regression we are trying to account for the variance in an outcome or dependent variable $(Y)$ by examining the influence of a number of independent, or predictor, variables.

The overall process of multiple regression is to analyze a set or group of several predictor variables with a single outcome variable. The overall $R^2$ is the "multiple coefficient of determination" that refers to the amount of variance in the outcome variable that is accounted for by the combined set of predictor variables. This figure can be tested for significance in order to determine whether the overall combined model is likely to be 0 in the population. By using multiple regression as in building a model, we can try to increase the $R^2$ value by adding variables of interest, and examine the specific contributions of individual predictors or groups of predictors.

Multiple regression is a technique that evaluators can use in a variety of ways and for different purposes. In the following sections we will discuss the ways in which multiple regression results can be interpreted in order to aid in understanding a specific problem. As a method for modeling, I will briefly describe the use of control and mediation variables before we explore the interpretation of results.

## Control Variables

Even though correlation does not allow us to make causal statements, it does allow us other possibilities for discovery. As we have already seen with bivariate regression, we can use the correlation between two variables to help us better predict one variable when we know the value of the other. We can also use the properties of regression to "explain" variance in the sense that we can identify proportions of variance in a target (outcome, criterion) variable linked to a specific, or several, predictors. That is the rationale for the coefficient of determination ($R^2$).

We can gain additional information about the relationships among variables when we "add" variables to a regression analysis. Often, when we add variables, we discover that our original regression results (between a predictor and an outcome variable) increase or decrease. This might be for many reasons, one of which is that the new variable shares a considerable amount of information with the existing predictor variable. If so, we would then gain further insight into our original regression relationship. Perhaps the original bivariate relationship is really the function of a third variable that is related to both our predictor and outcome variables.

When we make these changes to a regression model, we introduce variables as controls. That is, evaluators add variables and hold their values constant to provide a clearer picture of how the initial predictor alone affects the outcome, without the influence of the added variable. In this and the remaining chapters, we will explore how to use variables as controls; that is, what is the impact on the bivariate relationship when another predictor is added? How does this additional variable act to change the variance explained in the outcome variable? A more formal understanding of "control" concerns the understanding of research design, which evaluators often use to assess program impact. In Chapter 2, under the section, "The Evaluator's Tools," I discussed briefly how evaluators might use quasi-experimental design to compare the outcomes of different program conditions (e.g., membership in an intervention group versus nonmembership, or membership in another intervention group). This is the broad condition of control. The evaluator is assessing the program outcomes by controlling or limiting influences outside the program participation (either being in a group or not) on the outcome. In effect the evaluator is assessing the influence of the different treatment groups on the outcome when all other influences (e.g., individual differences) are the same.

Of course, with quasi-experimental design the evaluator often does not have the luxury of assigning subjects to different groups randomly, which would help ensure that individual differences were equal across treatment groups, but must measure the results of subjects already in existing groups. In these situations the evaluator is doing whatever is possible to ascertain that the groups to be compared are as equivalent as possible and thus control their influence on the treatment–outcome relationship.

The primary way we will discuss controls in this book is in terms of influence. What is the influence of one predictor on an outcome when we have controlled the impact of another predictor (or multiple predictors) on the original relationship?

## Mediator Variables

The way I described control variables places primary importance on understanding the influence of predictors as they together explain the variance in an outcome. Using multiple regression in this way assumes that the evaluator has some reason for adding the variables s/he adds as additional explanation. The rationale may lie in variables suggested by previous research or simply in exploring what variables emerge in the evaluation setting. Typically the researcher is limited as to what data are available and may explore the impact of available data on the outcomes. This does not mean that the modeling is without a causal framework, but it does reflect a more exploratory aim for the research. So more importance is placed on prediction, that is, what works best to explain the outcome variance, than on formal explanation.

Multiple regression is a statistical procedure that also can be used in analyses that formally assess the causal direction of relationships among variables. Some statistical techniques, like path analysis and structural equation modeling,[19] are designed to confirm and test the causal relationships among specific variables identified through prior theoretical models. (We will discuss such techniques briefly in Chapter 11.) Using multiple regression methods in this way allows the researcher to identify the specific causal relationships among a set of variables.

---

[19]There are several good treatments of path analysis and structural equation modeling. A particularly useful resource, using the SPSS®-related program AMOS is provided by Byrne (2010).

Mediator variables are like control variables, in the sense that they play a role in identifying specific influences on an outcome variable. However, the primary use of the more formal techniques is to examine in detail what role each variable plays in the theoretical trajectory of the research, and what the relationships are to the other variables in the analysis. In short, control and mediator variables are similar in how they help to describe the impact of specific variables in the evaluation research, but mediator variables are more closely identified with the formal attempt to support or disconfirm a theoretical position.

An example related to our earlier discussion of workplace democracy concerns workers who are given the opportunity to participate in decision-making (workplace democracy) and so are posited to be more satisfied in their work (worker satisfaction) than those not given the opportunity. The relationship between these two variables may be due to another variable, however: their attitudes toward participation in the workplace (desire for participation). Therefore "desire for participation" may serve as a mediating variable between opportunity for participation and worker satisfaction, since the influence of a democracy–satisfaction relationship at least partially "flows through" the attitude variable. Figure 7.2 shows how this relationship might result from a formal assessment of the causal directions.

The solid lines in the figure indicate that workplace democracy has a "direct" relationship to worker satisfaction, but it has an "indirect" relationship as well, the one that is mediated by desire for participation. The strength of the relationship between workplace democracy and worker satisfaction might be enhanced or diminished as a result of adding desire for participation to the analysis. Assessing these impacts will allow the evaluator interested in causal modeling to specify the relationships among all the variables. If the workplace democracy–worker satisfaction relationship disappears entirely when the desire variable is added to the analysis, for example, it would indicate that the

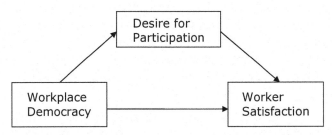

**Figure 7.2**  Diagram of a potential mediator relationship.

desire variable entirely mediates or carries the entire influence of the relationship. In this example it would mean that simply having a decision-making structure would have no impact on worker satisfaction unless the worker desires the opportunity for decision-making.[20]

The statistical techniques I mentioned earlier—path analysis and structural equation modeling—are available to analyze these mediation patterns formally. Cohen et al. (2003) also discuss mediating variables among other types of variables in causal analyses. I encourage you to seek out these sources, along with other similar references, if you want to dig a little deeper into the subject of mediating variables.

## USING MULTIPLE CORRELATION TO CONTROL VARIABLES: PARTIAL AND SEMIPARTIAL CORRELATION

Partial and semipartial "forms" of correlation provide the evaluator more information about the deeper patterns in data. *Partial correlation* can be useful for most of the situations we have just discussed, since it involves examining the relationship between a predictor and an outcome after the effects of another predictor are "taken out of" both the original predictor and outcome variables. *Semipartial*, or part correlation, is a process that allows us to identify the "unique" explanation of the original predictor to the outcome after the effects of another predictor are removed from the original predictor. The part correlation will be particularly useful when we discuss multiple regression because it represents a way to measure the unique contribution of a predictor (among multiple predictors) to the variance in the outcome variable.

In the discussions that follow I will make use of a common device in regression analysis, Venn diagrams. As shown in Figure 7.3, there are three circles. The top circle, labeled "$Y$" represents the outcome variable, whose variance we are trying to explain. Think of this entire circle as the "variance space" or total variance in the outcome variable. The bottom right circle, labeled "$X_1$," represents one predictor, while the bottom left circle, labeled "$X_2$," represents a second predictor.

The letters in the middle represent various "pieces" of variance in $Y$. Parts A and C represent unique contributions of $X_1$ and $X_2$, respectively, to the variance in $Y$. Part B represents the intercorrelation between the two predictors (the entire "football" or overlap of $X_1$ and $X_2$, not just

---

[20]The foundational resource for exploring mediation analysis is Baron and Kinney (1986). This resource is very helpful for identifying the conditions under which mediating influences are registered.

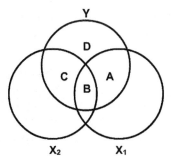

**Figure 7.3**  Venn diagram representing the relationship between predictors and outcome variable.

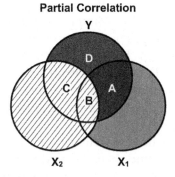

**Figure 7.4**   Isolating the effects of a predictor variable on an outcome through partial correlation.

the part of both that overlap only with $Y$). This is a difficult part of understanding the relationships, since it contains information that we cannot assign to either predictor unambiguously. Partial and semipartial correlations are different ways of understanding these relationships.

## PARTIAL CORRELATION

Partial correlation is a process whereby the researcher can assess the correlation between an outcome and one of the predictor variables ($X$) while holding the other predictor variable(s) constant. What this means, in practice, is that we are taking into account all of the various "pieces" of the overall set of correlations. When you add the third circle (representing a second predictor) several overlapping sections appear. It is clear that in Figure 7.4, both of the predictor variables are correlated with each other as well as each being correlated with $Y$. Partial correlation is a way of isolating the effects of a single predictor variable on an outcome variable, holding the other predictor variable(s) constant.

Partial correlation is technically the correlation between two variables from which the effects of one or more other variables have been removed. Figure 7.4 shows this as the diagonally marked portions C and B (the contribution of $X_2$) are removed from the relationship between $X_1$ and $Y$. The resulting relationship allows the researcher to measure what proportion of the remaining variance in $Y$ is attributable to $X_1$.

This correlation might best be understood through an example with the overall database we used earlier. Partial correlation might allow us to address the question, What is the relationship between math achievement and family income (F/R) when a predictor like ethnicity is taken out of both variables? In previous sections involving two variables, we explored the relationship between math achievement and family income (F/R). Suppose that we are curious about the effects of ethnicity as well as family income on math achievement. We can add this variable to the existing two variable relationship to see how much information it might add.

In the analyses that follow, we will use a variable to represent school-level ethnicity that is commonly used in educational research, "percent white" students per school.[21] This does not represent ethnicity as an individual characteristic, but it is one available indicator of the proportion of a school representing greater heterogeneity (i.e., lower scores on the variable) or greater homogeneity (i.e., higher values) among the schools' ethnic makeup.[22] We will use this generic indicator of *ethnicity* next. In a section below, we will make reference to findings based on different ethnic group populations (other than PercentWhite).[23]

---

[21] Different states indicate this ethnicity category differently. Some use "percent Caucasian" as a way of identifying this grouping.

[22] We could just as easily conceptualize this as "% Non-Caucasian" population at the school, since we are measuring the extent to which a school is not populated by Caucasians, but the method above is standard with researchers.

[23] Since we are using school-level data, no attribution can be made in the results about any individual student of any ethnic grouping. For example, if we perform an analysis concluding that there is an inverse relationship between school-level achievement and the proportion of a school's ethnic group, we cannot know which students in that ethnic group performed well, or which students performed poorly, in which schools. We can only conclude that there is a relationship between these variables among all the schools in our study. We are thus attempting to understand broad trends in student achievement.

Table 7.1  Correlation Matrix for Predictor and Outcome Variables

| | | Correlations | | |
|---|---|---|---|---|
| | | MathPercent MetStandard | PercentWhite | FreeorReduced PricedMeals |
| MathPercent MetStandard | Pearson Correlation | 1 | .480** | −.622** |
| | Sig. (2-tailed) | | .000 | .000 |
| | N | 1050 | 1050 | 1050 |
| PercentWhite | Pearson Correlation | .480** | 1 | −.652** |
| | Sig. (2-tailed) | .000 | | .000 |
| | N | 1050 | 1050 | 1050 |
| FreeorReduced PricedMeals | Pearson Correlation | −.622** | −.652** | 1 |
| | Sig. (2-tailed) | .000 | .000 | |
| | N | 1050 | 1050 | 1050 |

**Correlation is significant at the 0.01 level (2-tailed).

Table 7.1 (from specifying "Analyze–Correlation–Bivariate" in SPSS®) shows the zero order correlations among all three variables.

As is noted in Table 7.1, and in previous sections, the correlation between math achievement and F/R is −.622. The correlation between ethnicity and achievement is .480, indicating that, considered independently, ethnicity explains over 23% of the variance in achievement ($.48^2$) among the study schools. The inverse correlation between F/R and ethnicity is quite high at −.652, so we might question how these variables relate to one another.

SPSS® allows the user to conduct a partial correlation analysis by choosing "Correlation-Partial" after choosing the "Analyze" drop-down menu, as shown in Figure 7.5.

Figure 7.6 shows the submenu that allows you to specify a "control" variable (in the "Controlling for:" window) along with the bivariate correlation procedures. The partial correlation allows us to remove the contribution of FRLunch from both achievement and ethnicity so that we can get a better picture of what ethnicity contributes to school-level achievement. In this way, partial correlation "controls" for the influence of FRLunch on ethnicity and achievement.

Table 7.2 shows that F/R may be a mediator in the relationship between ethnicity and achievement. Recall that the zero-order

[24]The zero-order correlation is the bivariate relationship or simple correlation between two variables.

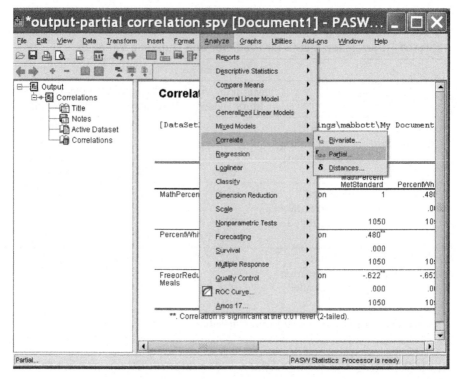

**Figure 7.5**  SPSS® menu specifying a partial correlation procedure.

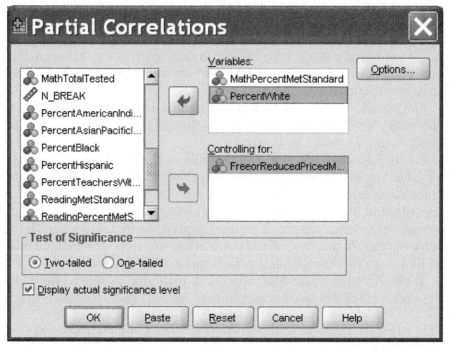

**Figure 7.6**  Submenu specifying variables for partial correlation.

**Table 7.2   Partial Correlation Results Controlling F/R**

Correlations

| Control Variables | | | MathPercent MetStandard | PercentWhite |
|---|---|---|---|---|
| FreeorReduced PricedMeals | MathPercent MetStandard | Correlation | 1.000 | .125 |
| | | Significance (2-tailed) | | .000 |
| | | df | 0 | 1047 |
| | PercentWhite | Correlation | .125 | 1.000 |
| | | Significance (2-tailed) | .000 | |
| | | df | 1047 | 0 |

correlation[24] between ethnicity and achievement was .48, indicating a significant positive relationship with a coefficient of determination of .23 ($.48^2$). Thus ethnicity alone accounts for 23% of the variance in math achievement. However, when we control for F/R by using the partial correlation, the correlation between ethnicity and achievement drops to .125 with a coefficient of determination of .016 (or about 2%), almost no relationship. Ethnicity has an impact on achievement, but primarily through the agency of F/R, which has a powerful relationship to achievement.

There is a danger in some studies' usage of partial correlation where there is a great deal of intercorrelation between the predictors. As Pedhazur (1997, pp. 171 ff) suggests, partial correlation can often remove too much information in some studies. To take our example above, we removed the influence of FRLunch on ethnicity and achievement through partial correlation. However, doing so could have removed too much of the information in "ethnicity" beyond that contained in FRLunch. When we are dealing with a limited number of variables, we often may make these limitations. This is also an important lesson for making sure that we have a causal framework for helping us determine the nature of the influences among the variables.

## SEMIPARTIAL (PART) CORRELATION

We come now to a very important measure for assisting in the interpretation of multiple regression findings. *Semipartial, or part, correlation* is the relationship of a predictor to an outcome when the effects of a third variable are taken out of the predictor variable but not

**Semipartial Correlation**

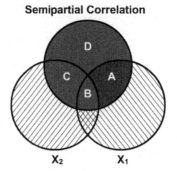

**Figure 7.7** Isolating the effects of a predictor variable on an outcome through part or semipartial correlation.

the outcome variable. As shown in Figure 7.7, the effects of the "third variable" $X_2$ are taken out of the predictor variable $X_1$ (i.e., "removing" section B and leaving only section A) but not out of the outcome variable (leaving sections C and B with section D). What remains is the correlation between the "unaffected" portion of the predictor and the outcome variable. In terms of the diagram, the part correlation is the correlation between (A) and the entire outcome variable y.

*When the part correlation is squared*, the resulting figure represents the proportion of the variance in the outcome variable taken into account by the unaffected predictor variable. In other words, the squared part correlation represents the *"unique" influence* of a predictor upon an outcome when other variables are controlled.

We can use the data set we introduced in the last section for an example of part correlation. SPSS® does not provide part correlation through the convenience of a separate menu as it does for partial correlation, so we must calculate it directly from the data. There are a number of algorithms for this purpose, but the easiest method is using the SPSS® output for multiple regression where you can request part correlations. We will wait until we get to that section to review how to obtain the part correlation results using the SPSS® procedures.

If we jump ahead to a preview of the SPSS® results, we will find that the part correlation between F/R (predictor) and math achievement (outcome) controlling for ethnicity is −.41. Recall that the zero-order correlation between F/R and achievement was −.622. Thus removing the effects of ethnicity from F/R lowers the zero-order correlation, but the resulting figure is still strong and significant. By squaring the part correlation, we can see that F/R makes a *unique contribution* to the variance in achievement of approximately 17% ($-.41^2$).

## DISCOVERY LEARNING

While correlation is helpful in illuminating the relationship between two variables, we also know that a complex outcome variable may involve more than one predictor. Multiple regression can help evaluators understand how several predictors analyzed at the same time can account for the variance in an outcome variable.

Substantively, it is intuitively obvious that math achievement is due to more than family income at a school level, and that worker satisfaction is an outcome of more than simply having the ability to participate in decision-making. A more comprehensive attempt to understand either of these would include incorporating additional predictor variables into the models.

When a third variable is "added" to a two-variable relationship, there are several ways in which the original relationship may change. It is important to look at all the interrelationships together in order to see what new information the additional variable can contribute. Partial and part correlation are two procedures that can be very helpful for understanding the nuances in these interrelationships.

When used alone, the proportion of a school whose students are classified as lower income is highly related (inversely) to the school-level math achievement. By using partial correlation, we find that school-based ethnicity is related to math achievement primarily through its relationship to the school's lower income proportion in this set of data. Part correlation helps the researcher to understand the unique influence of a predictor variable on an outcome when additional predictors are held constant. The part correlation of a school's proportion of low-income students to math achievement is a very strong figure in this dataset, even after controlling for the school's ethnic proportion.

### Terms and Concepts

**Control variables**    Adding variables to a correlation or regression analysis to discover that our original regression results (between a predictor and an outcome variable) may increase or decrease.

**Mediator variables**    Variables similar to control variables but often designated as mediators in more structured analyses in which there are formal attempts to support or disconfirm a theoretical position.

**Multiple correlation**   Correlation of one outcome variable with two or more predictor variables in order to understand the relationships among the variables.

**Multiple regression**   Process of analyzing a set or group of several predictor variables with a single outcome variable.

**Partial correlation**   Examining the relationship between a predictor and an outcome when the effects of another predictor are "taken out of" *both* the original predictor and outcome variables.

**Path analysis**   A multivariate analysis where the attempt is to visually identify and quantify the nature and direction ("paths") of causal influence among the variables. These analyses are "driven" by theoretical assumptions and models and represented graphically.

**Quasi-experimental design**   An experimental design in which the experimenter does not have control of all potentially confounding variables that may impact the relationship between the "treatment" and the outcome. Evaluators most often do not have the ability to use probability designs to ensure comparability of treatment groups prior to the introduction of a treatment, and therefore the results cannot express only the impact of the treatment. Random sampling is often not an option to evaluators. In many cases, especially in educational research, evaluators deal with "intact groups" or groups that are already chosen and active, like school classrooms. In these cases the evaluator must rely on comparing treatment groups while controlling the extraneous variables as completely as possible, but realize that causal conclusions in these situations is not realistic.

**Semi-partial correlation (part correlation)**   A process that allows the evaluator to identify the "unique" explanation of the original predictor to the outcome when the effects of another predictor are removed from the original predictor.

**Structural equation modeling (SEM)**   A multivariate procedure making causal assertions based on theoretical frameworks that involve a series of models encompassing "structural equations" or multiple regression analyses of the data. The variables involved, being specified from theory, may be unobservable ("latent") and either "exogenous" (independent influences) or "endogenous" as well as observable. SEM uses visual models to assist with the adjustment of data to arrive at a model that best explains the relationships among the data.

## Real World Lab—Partial and Semipartial Correlation

We discussed partial correlations in this chapter using the school-level data correlating reading and math achievement with family income and ethnicity. The database below is a sample of the data that contains these variables. Use these data to respond to the questions that follow. We will wait until the next chapter (multiple regression) to discuss the results of the semipartial correlation analyses.

| MathPercent MetStandard | ReadingPercent MetStandard | PercentWhite | FreeorReduced PricedMeals |
|---|---|---|---|
| 31 | 50 | 89 | 2 |
| 84 | 96 | 72 | 14 |
| 33 | 53 | 15 | 65 |
| 70 | 81 | 41 | 42 |
| 72 | 97 | 92 | 15 |
| 51 | 80 | 67 | 23 |
| 37 | 61 | 11 | 89 |
| 28 | 56 | 52 | 87 |
| 36 | 64 | 93 | 24 |
| 70 | 93 | 94 | 10 |
| 60 | 90 | 84 | 13 |
| 50 | 78 | 76 | 20 |
| 69 | 91 | 85 | 17 |
| 88 | 100 | 74 | 3 |
| 90 | 90 | 93 | 49 |
| 73 | 90 | 80 | 19 |
| 35 | 78 | 18 | 84 |
| 49 | 69 | 62 | 74 |
| 63 | 92 | 68 | 26 |
| 65 | 92 | 92 | 42 |
| 88 | 96 | 80 | 1 |
| 54 | 86 | 72 | 39 |
| 43 | 80 | 82 | 17 |
| 63 | 84 | 81 | 46 |
| 57 | 83 | 78 | 18 |
| 55 | 84 | 73 | 17 |
| 62 | 91 | 94 | 33 |
| 75 | 87 | 74 | 32 |
| 62 | 91 | 85 | 19 |
| 89 | 96 | 73 | 49 |

Create a correlation matrix and examine the relationships among the study variables:

1. What are the results of the bivariate correlations? Do they match the analyses discussed in the partial correlation section?
2. Conduct partial correlation analyses on the variables as we did in the chapter discussion. What is the impact on the relationship between reading achievement and school-level ethnicity when family income is removed from both of the study variables?
3. Interpret the results of your analyses. What do the results of the partial correlation analyses indicate?

# 8

# MULTIPLE REGRESSION

As we noted earlier, multiple regression is used to *predict* values on an outcome variable using two or more predictor variables, and to help *explain* the relationship among study variables according to a theoretical framework. The formula for multiple regression simply adds a series of predictor variables to the linear regression formula we examined earlier, with a beta coefficient (slope) for each. The beta coefficients tell us what changes occur in the outcome variable ($Y$) as a result of a predictor variable $X$, when other predictors are controlled, or held constant. The beta coefficients of each predictor variable can be tested to determine whether they are significant predictors of the outcome when the additional predictors are controlled.

You can obtain a measure of the accuracy of the multiple prediction of the outcome variable ($Y$) by the use of confidence intervals. Like the process used for single variable regression, you would first calculate the "standard error of multiple estimate." Then you could use the $Z$ values to help you establish the limits within which the actual value of $Y$ would be likely to fall, within a certain level of confidence.

The multiple regression procedure uses a computer-based statistical analysis program like SPSS® to add predictor variables sequentially to an analysis to detect changes in the $R^2$ figure. Through this method, one can observe the impact of specific predictor variables on the explained

*The Program Evaluation Prism: Using Statistical Methods to Discover Patterns,*
by Martin Lee Abbott
Copyright © 2010 John Wiley & Sons, Inc.

variance in the outcome variable. Really, this is the best way to assess "effect size" with $R^2$. We will discuss different ways of adding predictor variables to a multiple regression analysis in a later section in this chapter. The results of the analysis will vary depending on which method you choose.

In other readings you will learn about methods similar to regression analysis, like path analysis and structural equation modeling, to name just a very few. We will provide a brief description of these techniques in the last chapter of this book.

In what follows we will look at a multiple regression procedure using two predictor variables. This will extend what we learned earlier in examining a one-variable model, and it will lay the groundwork for examining procedures that add more predictor variables. There are also some special features in multiple regression when we use predictors that are categorical in nature (not continuous, but variables male/female, treatment/control, etc.) that we will examine in later sections. For now let's explore the data set we have been using to see how multiple regression can take the information we gained through multiple correlation to better predict an outcome variable.

As I stated in the Preface, any book dealing with advanced statistical procedures is done with reference to others that have helped advance the thinking about the subject. The treatment of multiple regression topics that follows has benefited greatly from the work done by such authorities as Pedhazur (1997), Cohen et al. (2003), and Tabachnick and Fidell (2007). I encourage you to seek out these works as you continue to grow in your interest and reliance on multiple regression topics.

## MULTIPLE REGRESSION WITH TWO PREDICTOR VARIABLES

### Uses of Multiple Regression

As I mentioned in an earlier chapter, no one variable alone will account for all the variance of another. Thus we might add additional predictor variables to the original two-variable model and see what overall explanation it provides. By looking at all the relationships at the same time, we can get a better picture of how each predictor contributes to the outcome when all the other predictors are controlled.

If we "regress" an outcome variable on a set or group of several predictor variables, SPSS® will provide a measure of how much variance in the outcome variable is accounted for by the combined set of predictors. This is $R^2$, the "multiple coefficient of determination" that we discussed earlier.

## Multiple Regression Outcomes

Multiple regression can also be used to create a more accurate prediction of outcome values, since we are adding information by adding predictors. We can monitor the changes in the $R^2$ as we add predictors to see whether or not they add significant information or explanations of the outcome by increasing $R^2$.

From the overall process we observe two sets of findings:

1. *The omnibus test.* This is the significance test of the $R^2$. As we will see, SPSS® provides a number of indicators of this test. This determines whether the regression coefficients of the predictors is nonzero. (It is possible for the omnibus test to be significant with "marginal" predictors.)

2. *Tests of significance for individual predictors.* The beta coefficients provided by SPSS® indicate what changes occur in the outcome as a result of a change in a predictor variable, when the other predictors in the equation are controlled or held constant. Each predictor is accompanied by a statistical test ($t$ test) that determines whether each slope is nonzero.

Beyond these two critical findings, SPSS® output provides a great deal of information that allows us to gain further insight about the predictors. We will discuss several important features of the predictors by examining the SPSS® output from the database we used earlier.

The SPSS® procedure for obtaining a multiple regression starts with the "analyze" menu, with a selection of "regression–linear." This will provide the window shown in Figure 8.1. At this window, you will simply place the outcome variable in the "dependent" window, and the predictors in the "independent(s)" window. There are several things to notice about this window, but for now, we can simply choose "OK," and we will get a standard regression output. We will return to discuss some of the other options as we make reference to them.

If we use the database we explored earlier and specified as our outcome variable, the percentage of students in the schools who met the state standard in math, we would obtain the output shown in Table 8.1. In this example our predictors are ethnicity and F/R as used earlier.

***Omnibus Findings for the Overall Model*** The top two panels of Table 8.1 provide information relevant to the omnibus test. In this example the $R^2$ of .397 (the shaded figure in the third column of the top panel) indicates that almost 40% of the variance in the outcome

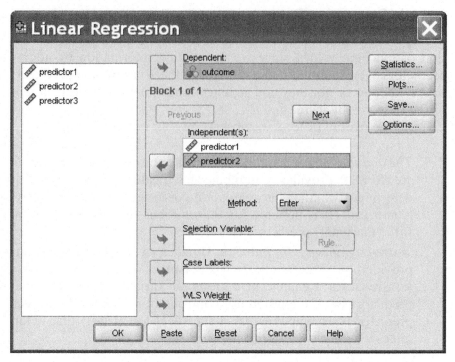

**Figure 8.1** SPSS® menu specifying linear regression.

variable, math achievement, is accounted for by the two predictors ethnicity and F/R. The multiple correlation is indicated in the second column of the top panel, $R = .630$. Notice also that the *adjusted R square* is reported. As I mentioned in the bivariate regression chapter, the adjusted $R^2$ is a value showing the likelihood of the $R^2$ value being obtained with all the data from the actual population in the analysis. The adjustment to $R$ takes into account sample size and number of predictors to provide a value that would more likely reflect a population-based analysis. Because we are using over 1000 cases, the $R$ squared and adjusted $R$ square figures are almost equal. The right-most column shows the standard error of the estimate, which, as we discussed above, is a measure of variation of the data points around the regression line, or an estimate of the accuracy of the prediction.

The middle panel ("ANOVA") shows the $F$ test based on the sums of squares and provides the significance test for $R^2$. As indicated in the final column, the omnibus test is statistically significant (see the shaded figure). This ANOVA analysis also provides the ability to calculate the $R^2$ directly by dividing two of the sums of squares figures. When we

**Table 8.1    SPSS® Output for Multiple Regression with Two Predictors**

Model Summary

| Model | R | R Square | Adjusted R Square | Std. Error of the Estimate |
|---|---|---|---|---|
| 1 | .630[a] | .397 | .396 | 13.307 |

a. Predictors: (Constant), PercentWhite, FreeorReducedPricedMeals.

ANOVA[b]

| Model | | Sum of Squares | df | Mean Square | F | Sig. |
|---|---|---|---|---|---|---|
| 1 | Regression | 122,068.162 | 2 | 61,034.081 | 344.669 | .000[a] |
| | Residual | 185,403.298 | 1,047 | 177.081 | | |
| | Total | 307,471.460 | 1,049 | | | |

a. Predictors: (Constant), PercentWhite, FreeorReducedPricedMeals.
b. Dependent Variable: MathPercentMetStandard.

Coefficients[a]

| Model | | Unstandardized Coefficients | | Standardized Coefficients | | |
|---|---|---|---|---|---|---|
| | | B | Std. Error | Beta | t | Sig. |
| 1 | (Constant) | 67.952 | 2.325 | | 29.232 | .000 |
| | PercentWhite | .095 | .023 | **.129** | 4.089 | .000 |
| | FreeorReduced PricedMeals | −.390 | .023 | **−.538** | −16.995 | .000 |

a. Dependent Variable: MathPercentMetStandard.

discussed residuals above, we noted that the amount of variance in the outcome explained by the predictors is known as "regression" and is shown in the second column of the middle panel (shaded) as the regression sum of squares. This sum of square figure can be divided by the total sum of square figure (also shaded) to yield the $R^2 = .397$ we saw above (122,068.162/307,471.460).[25]

***Individual Predictors***   The last panel of Table 8.1 provides the information for our second set of findings, the tests of significance for individual predictors. The last two rows of this panel provide information about the predictor's ethnicity and F/R. The right-most column (*t*-test results) shows that each of these is a significant predictor of math

[25]You can confirm this by calculating the proportion of the variation in the outcome due to error, or residual, and subtracting this from 1.0. In this example, 1 − (185, 403.298/307,471.46) = .397.

achievement when the other is held constant. Technically a significant finding indicates that the $b$ coefficient for an individual predictor is nonzero when controlling for the effects of the other predictor(s).

We should note, however, that statistical significance may be misleading when we are using a large database such as we have here. With over 1000 cases, even small relationships are easily considered statistically significant. By concentrating on *effect sizes*, we will be able to determine the absolute and relative effects of each of the predictors on the outcome. The statistical test for the individual predictors is provided by the $t$-test results in the next to last column. The last panel also provides information necessary for creating the regression equation. As we mentioned above, there are different ways of approaching the equation. If we want to concentrate on *standardized results*, we would use the standardized beta coefficients in the third to last column (shown in bold font), yielding an equation[26]

$$\hat{Z}_{(achievement)} = (.129)Z_{(ethnicity)} + (-.538)Z_{(F/R)}$$

If we wish to use *unstandardized* results, maintaining the original scale of the raw data, we would create an equation from the unstandardized coefficients in the third column (shown in bold font) as follows:

$$\hat{Y}_{(achievement)} = (.095)X_{(ethnicity)} + (-.390)X_{(F/R)} + (67.952)$$

As an application of this equation, suppose that we want to predict an achievement value for a particular school that did not report its results initially. If we know that the ethnicity and F/R measures of this particular school are 39% and 67%, respectively, we could predict the achievement level. Thus we would predict that such a school would have an achievement level of 45.53, or stated differently, we would predict that 45.53% of the students in that school would meet the state math achievement standard.

$$\text{Predicted achievement} = (.095)(39) + (-.39)(67) + 67.952 = 45.53$$

***Additional SPSS® Results*** To obtain the SPSS® results shown in Table 8.1, we specified a multiple regression with two predictors. Using the same menu ("Analyze–Regression–Linear"), we can obtain additional information that is very helpful to our analysis. In the upper

[26]Recall that when there is a single predictor, the standardized beta is equal to the correlation between the outcome and predictor variables.

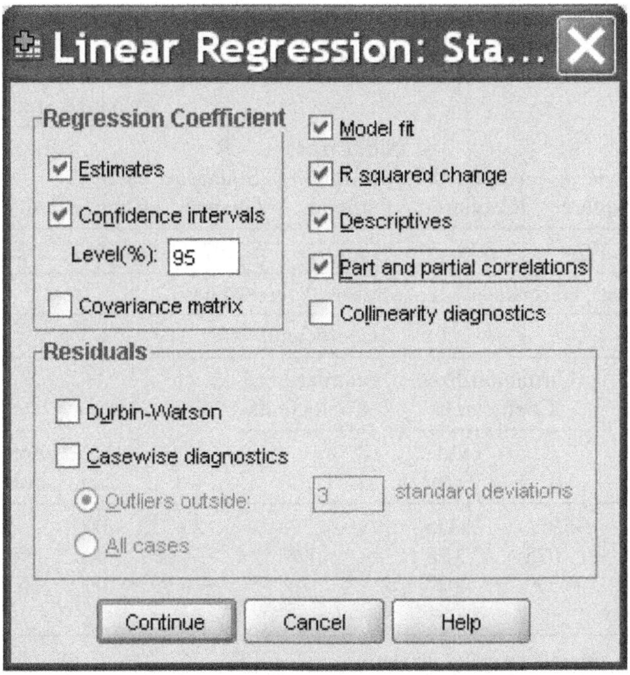

**Figure 8.2**  SPSS® menu specifying additional output values.

right portion of that panel is a tab titled "Statistics …" in which we can specify additional output. The screen shown in Figure 8.2 would result if we had selected the "Statistics …" tab on the previous menu.

"Model Fit" is already selected by SPSS® for the analyses we discussed earlier. We checked the other three boxes in this column, "$R$ squared change," "Descriptives," and "Part and partial correlations" because they will yield results that shed additional insight into the nature of the relationships among the variables in the analysis. By making these selections using our achievement database, we would obtain several additional panels of output. These include descriptive measures (means and standard deviations) of all the variables in addition to a correlation matrix showing the zero-order correlations.

The ANOVA panel would not change with these additional specifications so we will not report it here. The two panels of the output shown in Table 8.2, which we examined earlier in Table 8.1, have been expanded to include additional information.

The top panel of Table 8.2 adds "change statistics" indicating that the two predictors entered into the analysis together have changed the $R$ squared from 0 to .379 ("$R$ square change"). This change is

**Table 8.2 SPSS® Expanded Output for Multiple Regression with Two Predictors**

Model Summary

| Model | R | R Square | Adjusted R Square | Std. Error of the Estimate | Change Statistics | | | | |
|---|---|---|---|---|---|---|---|---|---|
| | | | | | R Square Change | F Change | df1 | df2 | Sig. F Change |
| 1 | .630[a] | .397 | .396 | 13.307 | .397 | 344.669 | 2 | 1047 | .000 |

a. Predictors: (Constant), FreeorReducedPricedMeals, PercentWhite.

Coefficients[a]

| Model | Unstandardized Coefficients | | Standardized Coefficients | t | Sig. | Correlations | | |
|---|---|---|---|---|---|---|---|---|
| | B | Std. Error | Beta | | | Zero-order | Partial | Part |
| 1 (Constant) | 67.952 | 2.325 | | 29.232 | .000 | | | |
| PercentWhite | .095 | .023 | .129 | 4.089 | .000 | .480 | .125 | .098 |
| FreeorReduced PricedMeals | −.390 | .023 | −.538 | −16.995 | .000 | −.622 | −.465 | −.408 |

a. Dependent Variable: MathPercentMetStandard.

significant, as indicated by the last column, "Sig. F. Change." Since both predictors were entered together as a group, we have one model that we are testing. This is indicated by the first column (model 1). If we had entered the predictors into the analysis in a different manner, we would be provided additional information.

The second panel shows three columns of correlation data under the final three columns ("Correlations"). The zero-order correlations are in the first of the three columns. The second column shows the partial correlations, which we discussed in an earlier section. The last column shows the part (or semi-partial) correlations. Recall earlier that we stated that squaring the part correlation results in the unique contribution of a predictor to the outcome variable while controlling for other predictors. Thus, for example, if you square (−.408), the part correlation between F/R and math achievement, you find that F/R *uniquely* accounts for 16.65% of achievement ($-.408^2$).

Take another look at the "Analyze–Regression–Linear" panel we discussed earlier. In the middle panel of Figure 8.3 is a drop-down menu entitled "Method." When we used the regression analysis, we simply accepted the default setting, which is "Enter," meaning that SPSS® will analyze all the predictors at the same time as they are entered by

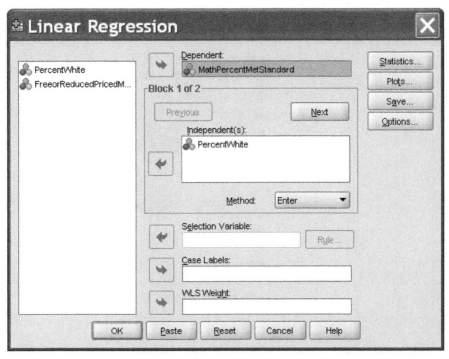

**Figure 8.3** SPSS® menu specifying different methods of entering predictor variables.

the user. We can glean additional information by how we enter the predictors, however. Below, we discuss several "entry methods" to show how different results are available depending on how the evaluator enters the predictors into the analysis. For now, we will focus on a common way of entering the predictors that will yield very helpful information.

Look at the panel in Figure 8.3 where we entered only the first predictor. That panel shows that we entered the ethnicity ("PercentWhite") variable by itself. We would see that this variable is considered its own "block" of variables by the title below the "Dependent" window entitled "Block 1 of 2," which indicates that whatever variables we entered at that time would constitute a group of predictors considered together. If, after we entered this variable we selected the tab entitled "Next," we would be able to enter the second predictor by itself. Figure 8.4 shows that we entered "FreeReducedPriceMeals" as the second predictor, and that it will constitute "Block 2 of 2." The user can toggle back and forth between the blocks by using the "Previous" and "Next" buttons.

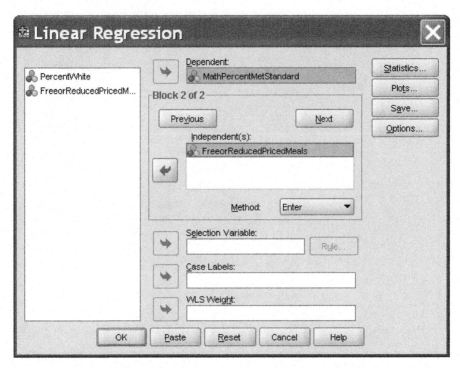

**Figure 8.4** SPSS® menu showing F/R added as block 2.

This method would result in two different models in the outcome (each containing the variables in the separate blocks.) The first model would simply be the one-variable regression between the first predictor and the outcome variable. The second model would *add* the second predictor and represent a two-predictor model. *The advantage in doing this is that you can observe what happens to the $R^2$ value when the second variable is added to the model.*

Using our achievement database as this example, we first entered ethnicity and then entered F/R in a separate block. The results are shown in the SPSS® output panels in Table 8.3.

The SPSS® outcome panels look similar to the ones we examined in earlier tables that included only a single predictor. However, since we added the predictors separately in the analysis, we now have two "models" that are reported in the same outcome panels. If you look at each panel of Table 8.3, you will see separate results reported for *model 1* and *model 2*. For example, model 1 reported $R^2$ as .231 in the top panel of Table 8.3. Since model 1 represents a regression analysis with only one predictor (ethnicity), the $R^2$ is simply the square of the

**Table 8.3    SPSS® Regression Output Created from Separate Entry of Predictors**

Model Summary

| Model | R | R Square | Adjusted R Square | Std. Error of the Estimate | Change Statistics | | | | |
|---|---|---|---|---|---|---|---|---|---|
| | | | | | R Square Change | F Change | df1 | df2 | Sig. F Change |
| 1 | .480[a] | .231 | .230 | 15.024 | .231 | 314.202 | 1 | 1048 | .000 |
| 2 | .630[b] | .397 | .396 | 13.307 | .166 | 288.838 | 1 | 1047 | .000 |

a. Predictors: (Constant), PercentWhite.
b. Predictors: (Constant), PercentWhite, FreeorReducedPricedMeals.

ANOVA[c]

| Model | | Sum of Squares | df | Mean Square | F | Sig. |
|---|---|---|---|---|---|---|
| 1 | Regression | 70,920.654 | 1 | 70,920.654 | 314.202 | .000[a] |
| | Residual | 236,550.805 | 1,048 | 225.716 | | |
| | Total | 307,471.460 | 1,049 | | | |
| 2 | Regression | 122,068.162 | 2 | 61,034.081 | 344.669 | .000[b] |
| | Residual | 185,403.298 | 1,047 | 177.081 | | |
| | Total | 307,471.460 | 1,049 | | | |

a. Predictors: (Constant), PercentWhite.
b. Predictors: (Constant), PercentWhite, FreeorReducedPricedMeals.
c. Dependent Variable: MathPercentMetStandard.

Coefficients[a]

| Model | | Unstandardized Coefficients | | Standardized Coefficients | t | Sig. | 95.0% Confidence Interval for B | | Correlations | | |
|---|---|---|---|---|---|---|---|---|---|---|---|
| | | B | Std. Error | Beta | | | Lower Bound | Upper Bound | Zero-order | Partial | Part |
| 1 | (Constant) | 34.530 | 1.399 | | 24.675 | .000 | 31.784 | 37.276 | | | |
| | PercentWhite | .351 | .020 | .480 | 17.726 | .000 | .313 | .390 | .480 | .480 | .480 |
| 2 | (Constant) | 67.952 | 2.325 | | 29.232 | .000 | 63.391 | 72.514 | | | |
| | PercentWhite | .095 | .023 | .129 | 4.089 | .000 | .049 | .140 | .480 | .125 | .098 |
| | FreeorReduced PricedMeals | −.390 | .023 | −.538 | −16.995 | .000 | −.435 | −.345 | −.622 | −.465 | −.408 |

a. Dependent Variable: MathPercentMetStandard.

zero-order correlation between ethnicity and achievement ($.48^2$). (You can also find this correlation reported in the "Zero-order" column of the bottom panel. Here the correlation between achievement and ethnicity is .48, which squared is .23.) This one predictor model is significant as shown by the results in the model 1 row of the top and middle panels (Model Summary and ANOVA panels, respectively). In both cases, the $F$ test is significant for the entire model. The model 1 row of the bottom panel (Coefficients) shows that the standardized beta coefficient is .480, which is the same as the zero-order correlation, according to our earlier discussions of one predictor models.

If we focus on *model 2 results*, we obtain some additional information.

- In the second row of the top panel, we see that $R^2$ is now .397, which has increased by .166 (.397 − .231). This increase is shown in the column titled "R Square Change," in the middle of the top panel, and it represents the additional information resulting from adding a second predictor (F/R).
- The final column to the right of this ("F Change") indicates that the change in the overall $F$ statistic as a result of adding the second variable is statistically significant. You can derive this same number (288.83) by squaring the $t$-test results reported in the middle of the bottom panel for the F/R predictor. In that case, $t = (-16.995)$, which, when squared, is 288.83.
- Model 2 results in the middle panel indicate that the omnibus results for the model with two predictors is significant.
- Results in the bottom panel (Coefficients) indicate that the regression coefficients (constant and beta coefficients) change when a second variable is added, which is to be expected. We still observe that the separate $t$-tests for the two predictors are significant, indicating that each is a significant predictor of achievement when the effects of the other are controlled.
- We also observe that the partial correlation for the first predictor (ethnicity) changes when the second predictor is added. As reported in the next to last column of the bottom panel under model 2, the partial correlation for ethnicity is .125. Thus the correlation between ethnicity and achievement is .125 when the contribution of F/R is taken out of both of them. (We observed these findings in our discussion of multiple correlation in Chapter 7, at Table 7.2.) The partial correlation for ethnicity when it was the only predictor was .48, the same value as the zero-order correlation, since there were no other variables to control. As we noted before, however, this drop indicates that the relationship between ethnicity and achievement may be mediated by F/R.
- The part (or semi-partial) correlation for F/R is (−.408), as indicated in the last column of the bottom panel under model 2. Recall that when we square this amount, we get the unique contribution of that predictor to the variation in the outcome variable. If we square (−.408), we get .166, which indicates that almost 17% of the

variance in achievement is *uniquely* accounted for by F/R. This figure (.166) is reported in the second row of the middle of the top panel ("R Square Change"). You might also recall that this same figure (−.408) emerged above (see Table 8.2) as the part correlation of F/R when we entered both predictors at the same time. Thus, regardless of how we enter the data, the part correlation helps us obtain the unique contribution to the variance of the outcome by a predictor when other predictors are held constant.

- Squaring the part correlation of subsequent variables added individually to a regression analysis will provide the amount of the change in the $R^2$ between models, and will be reported as "R Square Change." If more than one predictor is added to the same block, then all the predictors in the block will together represent the change in $R^2$ from the previous block. Regardless of what the "R Square Change" reports, the squared part correlation for an individual predictor represents the proportion of variance in the outcome accounted for by that one predictor, when the other predictors are held constant. For this reason you can create an equation for the squared part correlation as follows:

$$(\text{Part } r)^2 = R^2_{1.23} - R^2_{1.3}$$

- In this equation, which shows two predictors, the squared part correlation is equal to the difference between the $R^2$ when both predictors are included ($R^2_{1.23}$) minus the $R^2$ when only the other predictor is included ($R^2_{1.3}$). We noted in our discussion of multiple correlation in Chapter 7 that there are several algorithms for calculating part correlation directly, but using the regression output seems to be the most straightforward.

- When interpreting the direction of the relationships indicated by the squared part correlations, use the sign taken from the output table.

- The statistical test of the beta coefficient (*t* test) applies to the squared part correlation as well.

## MULTIPLE REGRESSION: HOW TO ENTER PREDICTORS

As we saw in the discussion above, the output for multiple regression can be very helpful for estimating the effects of individual predictor variables on the outcome variable when the additional predictors are controlled. We used a process in the second example where we entered the predictors in such a way as to show the changes on $R^2$ when we added predictors. This "hierarchical" method is common and allows the user to specify how variables should be entered based on a priori reasons.

*Hierarchical analyses* also allow you to enter sets of variables to see the effects on the $R^2$ according to the dictates of your theory/research question. When you arrive at the last "model" in the output, you can see what each contributes uniquely to the output.

In using the hierarchical approach, you will need to think about how to proceed based on the results. If a given predictor is not significant, should it be deleted from the analysis? For example, in the illustrative cases we discussed earlier, ethnicity was statistically significant, although it did not account for a great deal of the variance in math achievement. Was this because of the large database size, or because most of its effects were contained in F/R? These and other considerations would have a bearing on your decision to retain the predictor for further study or to delete it from the model.

Aside from hierarchical approaches, there are several other ways to "build" a regression model according to how the variables are entered into the analysis. Most of these methods are very specialized, and some are quite problematic if used indiscriminately. I will only mention a few here, but evaluators serious about the intricacies of using regression will want to consult additional sources on the subject.

Using our method of monitoring the $R^2$ as we add variables, we can see that the order by which we enter variables into the analysis will affect the results. If you look at the output in Table 8.3, you can see that entering ethnicity first will yield an $R^2$ change of .231 ($.48^2$), which represents the squared zero-order correlation because it is the first variable in the analysis. Now consider the output in Table 8.4 where we entered F/R first into the analysis.

As shown in the output in Table 8.4, the $R^2$ change attributable to ethnicity is now .01 (see "R Square Change" column under model 2 in the first panel of Table 8.4), since it follows F/R in the order of entry. Using the squared part correlation in the last column of model 2 in the bottom panel of Table 8.4 ($.098^2$) allows us to see what the unique effects are on the outcome variable (.0096). Changing the order of entry can have an effect on the interpretation, so *it is important to add variables to the model according to your considered logic prior to the analysis.* Yet note that the part correlations for both predictors are exactly the same regardless of the order of entry. Compare the final column of the Coefficients panels in Table 8.3 and Table 8.4.

**Table 8.4   SPSS® Regression Output Specifying F/R to Enter First**

Model Summary

| Model | R | R Square | Adjusted R Square | Std. Error of the Estimate | R Square Change | F Change | df1 | df2 | Sig. F Change |
|---|---|---|---|---|---|---|---|---|---|
| | | | | | Change Statistics | | | | |
| 1 | .622[a] | .387 | .387 | 13.407 | .387 | 662.676 | 1 | 1048 | .000 |
| 2 | .630[b] | .397 | .396 | 13.307 | .010 | 16.720 | 1 | 1047 | .000 |

a. Predictors: (Constant), FreeorReducedPricedMeals.
b. Predictors: (Constant), FreeorReducedPricedMeals, PercentWhite.

Coefficients[a]

| Model | | Unstandardized Coefficients B | Std. Error | Standardized Coefficients Beta | t | Sig. | Correlations Zero-order | Partial | Part |
|---|---|---|---|---|---|---|---|---|---|
| 1 | (Constant) | 76.822 | .842 | | 91.200 | .000 | | | |
| | FreeorReduced PricedMeals | −.451 | .018 | −.622 | −25.742 | .000 | −.622 | −.622 | −.622 |
| 2 | (Constant) | 67.952 | 2.325 | | 29.232 | .000 | | | |
| | FreeorReduced PricedMeals | −.390 | .023 | −.538 | −16.995 | .000 | −.622 | −.465 | −.408 |
| | PercentWhite | .095 | .023 | .129 | 4.089 | .000 | .480 | .125 | .098 |

a. Dependent Variable: MathPercentMetStandard.

## STEPWISE REGRESSION AND OTHER METHODS

The difficulty with order-of-entry schemes is that they are typically done by the program according to a set of "entry" and "removal" criteria that are not set according to your own rationale but according to statistical "impact." In *stepwise regression* methods, variables are chosen by the program to be entered into and removed from the model according to how large their impact is on the overall probability of the $F$ statistic. Thus the output could present results different from the output stemming from your own theoretically derived sequence of entering predictors.

To take a brief and simple example, the panel in Figure 8.5 represents the request for an analysis in which I specified three predictors for the outcome of math achievement: Percent White as the first, F/R as the second, and "Average Years of Educational Experience" as the third predictor. Assuming I am entering these variables according to my theoretical assumptions, I want to be able to understand how each

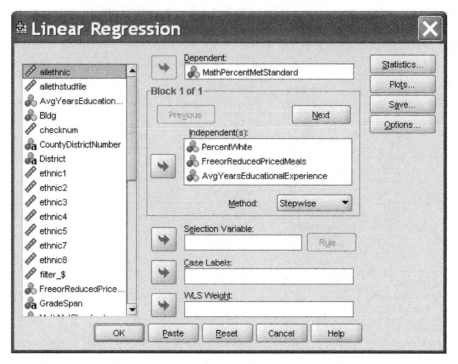

**Figure 8.5** SPSS® output specifying stepwise procedures.

figures in specifying the impact on the outcome variable math achievement. Figure 8.5 shows this specification along with the "Stepwise" procedure requested under the "Method" button in the middle of the panel.

When I choose stepwise procedures, the program makes decisions when to enter a predictor according to selection criteria. As you can see from Figure 8.6, the default "Stepping Method Criteria" specify adding variables to the analysis in steps if the probability of $F$ is at least .05. In subsequent steps each predictor is assessed with this same criterion as other predictors are added. If any of the variables exceed a probability of $F$ at .10 or more, that predictor will be excluded from the model.

In Table 8.5 I have reported the output for the analysis I discussed above using a sample of the database for illustration. In the first panel of Table 8.5, the Model Summary shows two models, the second of which includes two predictors. The ethnicity predictor is not included in the final models. The second panel of Table 8.5 shows the Coefficients summary, again excluding ethnicity. The third panel of Table 8.5 is the

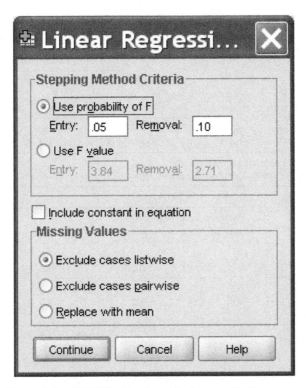

**Figure 8.6** SPSS® stepping criteria specification menu.

table that describes Excluded Variables and shows the entry–exit decisions made by the program.

This third panel shows that for model 1, both ethnicity and experience were excluded from the analysis in which F/R was retained on the basis of its higher probability of F value. In model 1, both ethnicity and experience are significant ($p < .024$ and $p < .008$, respectively). However, experience has a higher significance of $F$ value and is therefore included in the next step along with F/R. The $t^2$ value for experience ($2.667^2 = 7.111$, which is "F Change") qualifies it for entry, since it did not exceed the removal criterion (.10).

In model 2, the final model, experience is thus included, but ethnicity is now excluded, since it slipped above the .10$F$ criterion even though it was a significant predictor in model 1. It was not entered into the final model because it was slightly higher on the removal criterion than experience.

This example points out an important reminder when evaluators consider stepwise regression procedures. If a predictor does not meet

**Table 8.5   SPSS® Output for Stepwise Procedure**

Model Summary

| Model | R | R Square | Adjusted R Square | Std. Error of the Estimate | R Square Change | F Change | df1 | df2 | Sig. F Change |
|---|---|---|---|---|---|---|---|---|---|
| | | | | | Change Statistics | | | | |
| 1 | .632[a] | .399 | .398 | 13.045 | .399 | 389.082 | 1 | 585 | .000 |
| 2 | .638[b] | .407 | .405 | 12.978 | .007 | 7.111 | 1 | 584 | .008 |

a. Predictors: (Constant), FreeorReducedPricedMeals.
b. Predictors: (Constant), FreeorReducedPricedMeals, AvgYearsEducationalExperience.

Coefficients[a]

| Model | | B | Std. Error | Beta | t | Sig. | Zero-order | Partial | Part |
|---|---|---|---|---|---|---|---|---|---|
| | | Unstandardized Coefficients | | Standardized Coefficients | | | Correlations | | |
| 1 | (Constant) | 77.523 | 1.101 | | 70.401 | .000 | | | |
| | FreeorReduced PricedMeals | −.454 | .023 | −.632 | 19.725 | .000 | −.632 | −.632 | −.632 |
| 2 | (Constant) | 70.489 | 2.856 | | 24.677 | .000 | | | |
| | FreeorReduced PricedMeals | −.449 | .023 | −.626 | 19.586 | .000 | −.632 | −.630 | −.624 |
| | AvgYearsEducational Experience | .499 | .187 | .085 | 2.667 | .008 | .130 | .110 | .085 |

a. Dependent Variable: MathPercentMetStandard.

Excluded Variables[c]

| Model | | Beta In | t | Sig. | Partial Correlation | Tolerance |
|---|---|---|---|---|---|---|
| | | | | | | Collinearity Statistics |
| 1 | PercentWhite | .094[a] | 2.256 | .024 | .093 | .588 |
| | AvgYearsEducationalExperience | .085[a] | 2.667 | .008 | .110 | .995 |
| 2 | PercentWhite | .070[b] | 1.643 | .101 | .068 | .551 |

a. Predictors in the Model: (Constant), FreeorReducedPricedMeals.
b. Predictors in the Model: (Constant), FreeorReducedPricedMeals, AvgYearsEducationalExperience.
c. Dependent Variable: MathPercentMetStandard.

the criteria for entry, it is not included in the final model, *regardles of a theoretical rationale for including it*. This is evident in the example. While ethnicity was (just) beyond the removal criterion, it nevertheless was excluded. However, ethnicity might be a crucial variable to retain in the analysis if we want to understand the specific nature of all the relationships in the study, and it could have been a critical predictor implicated in prior research or theoretical consideration.

*Backward* and *Forward* are other types of entry methods available in SPSS®, but I recommend that you understand these very well before you use them and be able to defend *why* you chose them.

## ASSUMPTIONS OF MULTIPLE REGRESSION

The assumptions of linear regression with multiple predictors are largely the same as those for one predictor, as discussed in the chapter on bivariate regression. Obviously, since there will be more than one predictor, the assumptions would account for the differences of adding other predictors. For example, distribution of the residuals will reflect all the variables in the analysis, not just a single predictor, so this assumption takes into account the multiple predictors. In the same way, the other assumptions should be viewed with multiple predictors in mind.

### Multicollinearity

Mulicollinearity can be a serious problem for regression methods because it represents the extent of the intercorrelation among the predictor variables. If the intercorrelation is high, it makes it difficult to identify what impact each predictor has on the outcome when the others are held constant. If you refer back to the section on part and partial correlation, you will recall that methods exist that help us identify various "pieces" of the relationship among a set of variables. When intercorrelation is high, the overlap among the predictors may be so great as to make the overall analysis murky. Specifically, the confidence intervals for the regression estimates (betas) will be inflated with large intercorrelations, since the intercorrelations must be factored into the equation for deriving the estimates.

One remedy for multicollinearity is to examine the intercorrelations among a set of predictors before a regression analysis. If they are high, you might consider ways to simplify the analysis, making reference to your theoretical framework. Thus, for example, if two variables measure similar constructs, one may serve in your analysis rather than both. Unfortunately, there is no agreement about what constitutes "high," so you need to examine the correlations carefully. SPSS® includes a couple of diagnostic measures that help identify multicollinearity.

You can obtain these through specifying "Collinearity diagnostics" in the "Analyze–Regression–Linear–Statistics" panels. The example in Figure 8.2 did not include these measures, but you can see in Figure 8.7 that they can be obtained by checking the bottom box in the second column of choices ("Collinearity diagnostics").

If we had specified this box in our last analysis, we would have received the SPSS® output shown in Table 8.6. The last two columns

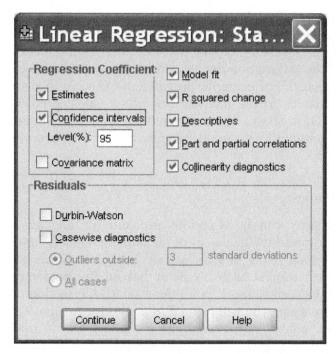

**Figure 8.7**  Specifying collinearity diagnostics for the regression analysis.

**Table 8.6  SPSS® Output Showing Collinearity Results**

| | | | | | | | | | | |
|---|---|---|---|---|---|---|---|---|---|---|
| | | | | | Coefficients[a] | | | | | |
| | Unstandardized Coefficients | | Standardized Coefficients | | | | Correlations | | Collinearity Statistics | |
| Model | B | Std. Error | Beta | t | Sig. | Zero-order | Partial | Part | Tolerance | VIF |
| 1  (Constant) | 67.952 | 2.325 | | 29.232 | .000 | | | | | |
| FreeorReduced PricedMeals | −.390 | .023 | −.538 | −16.995 | .000 | −.622 | −.465 | −.408 | .575 | 1.740 |
| PercentWhite | .095 | .023 | .129 | 4.089 | .000 | .480 | .125 | .098 | .575 | 1.740 |

a. Dependent Variable: MathPercentMetStandard.

list values for "Tolerance" and "VIF" for each predictor. These indicate the extent of intercorrelation between each predictor and the remaining predictors. "Tolerance" represents the extent of the intercorrelation of the predictors, obtained by subtracting the squared multiple correlation of the additional predictors with the predictor in question from 1. So, to use this example, if you review the data we noted earlier, the correlation between the two predictors F/R and ethnicity is (−.652). Thus Tolerance equals $1 - (-.652)^2$ or .575, the amount shown in the Tolerance panel in the output.

In this example there was only one additional predictor, so the Tolerance value can be easily calculated. When there are more than two predictors, you would use SPSS® to obtain the squared multiple correlation (by regressing the target predictor in question on the additional predictors) and subtract this value from one to obtain Tolerance for that target predictor.

As you can see, the *lower the Tolerance value*, the greater is the intercorrelation, and therefore the greater the likelihood of complications to the overall regression analysis due to multicollinearity. As I noted earlier, there is no good rule for how large the Tolerance value would be before you might be concerned about multicollinearity. For example, if you chose $r = .70$ between a target predictor and a second predictor (or set of other predictors) as a rule-of-thumb for a cutoff value, the resulting Tolerance value would be approximately .51 (or $1 - (.70)^2$). This would be a pretty strict cutoff—I would not be concerned at this level. But, if you chose $r = .80$, then the resulting Tolerance would be .36, which might concern me a bit more. If you think about it, this would be saying that only a small percentage (just over one-third) of the variance is left unaccounted for when two predictor variables are this highly correlated. This Tolerance Level will make a regression analysis murkier, since one of the things we are attempting in regression is to show the unique effects of predictors on outcome variables. I hope you learn to use Tolerance statistics to understand the extent of the intercorrelation among predictors.

While there are no universally accepted guidelines for Tolerance values, Cohen et al. (2003) suggest that values of .10 or less can indicate multicollinearity problems. Then again, SPSS® uses quite a low value (0.0001) to identify extreme collinearity.

*VIF* refers to a variance inflation factor, and this is another calculation that measures the impact of the intercorrelation of predictor variables. If the value is high, it indicates that the intercorrelations have inflated the variance of the regression estimates. VIF is simply 1/Tolerance, or in this instance shown in the output above, 1.74 (1/.575).

As with Tolerance, there is no generally accepted benchmark for what is considered high. Just remember what it is and how it is measured so that you can decide how to use it in your overall regression analysis. Cohen et al. (2003) note that VIF figures of 10 are quite high for most social science research, so that may be one benchmark to keep in mind as you examine your output.

## CLEANING THE DATABASE

I should add at this point that the researcher should use the procedures we discussed earlier to locate extreme scores that may affect the analysis. Remember that you do this by selecting "Save" on the panel for Linear Regression, whereby you can choose a variety of measures for detecting outliers. We discussed several above, but you can explore the others provided in SPSS® if you are concerned about your database.

We have explored a database with a very large number of cases as examples of how to proceed with a multiple regression analysis. It is unlikely that deleting a few (even several) problematic scores will substantially change the outcomes of the analysis. However, it is a good exercise to take the step of looking at the data carefully before you begin an extensive regression analysis.

## MULTIPLE REGRESSION WITH MORE THAN TWO PREDICTOR VARIABLES: RESEARCH EXAMPLES

I used two predictors in the sections above to demonstrate how to interpret regression analyses. However, the power of this method is that it can be used with many predictors. The results of regression analyses allow the evaluator to isolate the effects of a given predictor on an outcome variable while holding the effects of the remaining predictors constant. In this section I demonstrate how to use regression analysis with more than two predictors in some actual research projects.

### Predicting the Impact of School Variables on Teaching and Learning: The TAGLIT Data

We discussed TAGLIT data through a research example in Chapter 4 on correlation. The 2003 TAGLIT measures were based on a large number of respondents from different students and teachers across the country; the final database also included the responses of school administrators regarding objective school measures. As I described in the earlier analyses, I used FA and PCA to create four indexes from the teachers' responses that reflected their opinions and attitudes about several aspects teaching and learning, and the availability of technology support:

- Technology skill levels of teachers and students in elementary and secondary schools ("skill").
- Technology effect on classroom learning ("impact").
- Teachers' access to technology ("access").
- Teachers' assessment of available technology support ("support").

Evaluators had additional data available that could be coupled with the student and teacher responses. School administrators provided important data that could be used in a broader assessment of what might impact teaching and learning. One administrator from each of the schools that returned teacher and student responses responded to the TAGLIT instrument that assessed such things as the school planning process, availability of technology hardware and software, technical support, and technology revenue and expenditures. Evaluators also had access to information about the schools themselves, including ethnicity, students eligible for free or reduced price lunch, and urban/ rural status, among other descriptors.

Combining these sources of data involved the creation of school-based measures. To the existing school data descriptions (e.g., from school administrators), I aggregated the index scores from teachers to the school level, using the process I described in Chapter 3 on SPSS®. I used data from schools in which at least two MH teachers responded to the TAGLIT instrument. This way all the schools had a series of objective descriptors and the combined measures of teacher attitudes and opinions.

Figure 8.8 shows the SPSS® panel in which I specified the regression analysis. There are five predictors for this analysis, even though the small window in the center of the panel (under "Independents") does not show them all. Choosing the "Statistics" button in the upper right corner of the panel allows me to specify further the nature of the analysis I want with these data.

Figure 8.9 shows that I chose all the elements of the analyses that I have described in earlier sections of the book. When I run this analysis, SPSS® generates an output file with several components of the results. I obtain descriptive statistics (means and standard deviations) on all variables, a correlation matrix, and several tables showing the results.

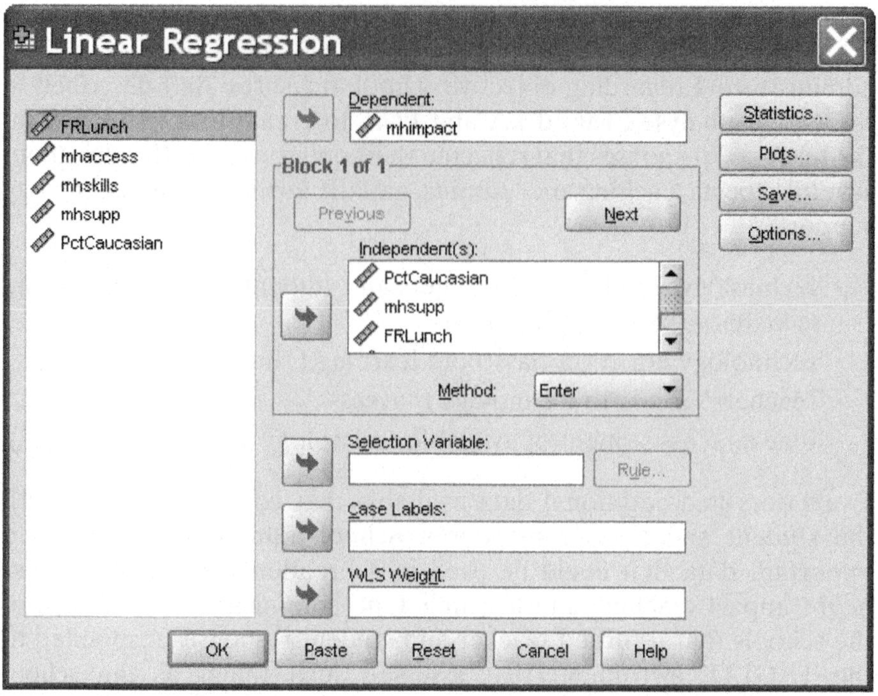

**Figure 8.8** SPSS® screen showing the MLR specification with more than one predictor.

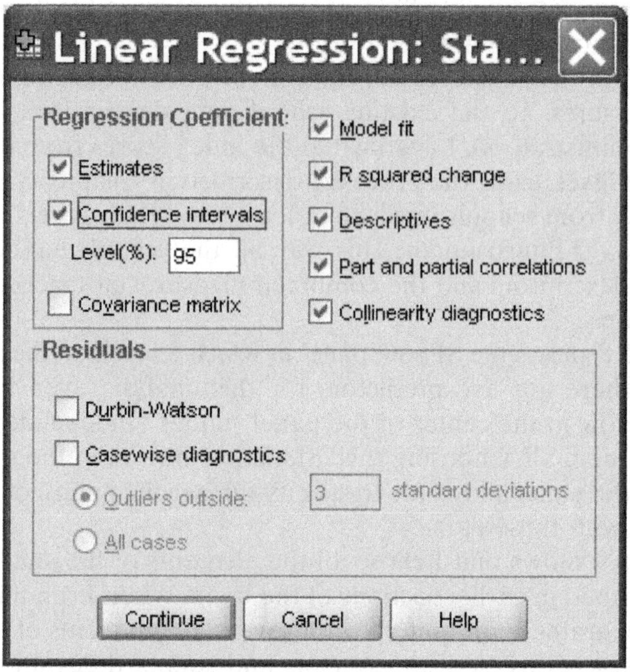

**Figure 8.9** SPSS® screen showing the requested statistics procedures used in the MLR analysis with more than one predictor.

***Omnibus Findings*** Table 8.7 shows the omnibus output from the SPSS® output file. As these panels indicate, this is a significant model ($F = 228.984$, p < .001) with an adjusted $R^2$ of .31. Thus the five predictors significantly predict the impact index, and together account for about 31% of the variance in the index. The school-based middle/high teachers' opinions regarding teaching and learning in the classroom is predicted by the aggregated teachers indexes of technology skills, access, and support, along with the percentage of students eligible for free/reduced lunch and the percentage of Caucasian students. We cannot make statements about individual teacher responses, but these results do indicate some important relationships among the study variables.

***Results of Individual Predictors*** Table 8.8 shows a series of results based on the individual predictors. Each of the predictors is significant, indicated under the "Sig." column. With a database this large, it is typical to have statistically significant results, which is the reason I like to focus on the effect sizes (squared part correlations) for the individual predictors. The results for FRLunch and PctCaucasian are interesting. The inverse findings for ethnicity and the positive influence of FRLunch are not typical in studies predicting achievement, but in this analysis there appear to be slight effects of schools with higher non-Caucasian populations and lower proportions of students eligible for lunch funding to show greater impacts of technology on teaching and learning. Again,

**Table 8.7   SPSS® Output for Omnibus Results**

| Model Summary | | | | | | | | | |
|---|---|---|---|---|---|---|---|---|---|
| | | | | | | Change Statistics | | | |
| Model | R | R Square | Adjusted R Square | Std. Error of the Estimate | R Square Change | F Change | df1 | df2 | Sig. F Change |
| 1 | .558ᵃ | .311 | .310 | .22990 | .311 | 228.948 | 5 | 2532 | .000 |

a. Predictors: (Constant), mhaccess, FRLunch, PctCaucasian, mhskills, mhsupp.

| ANOVAᵇ | | | | | |
|---|---|---|---|---|---|
| Model | | Sum of Squares | df | Mean Square | F | Sig. |
| 1 | Regression | 60.504 | 5 | 12.101 | 228.948 | .000ᵃ |
| | Residual | 133.826 | 2532 | .053 | | |
| | Total | 194.330 | 2537 | | | |

a. Predictors: (Constant), mhaccess, FRLunch, PctCaucasian, mhskills, mhsupp.
b. Dependent Variable: mhimpact.

**Table 8.8 SPSS® Output for Individual Predictor Results**

Coefficients[a]

| Model | Unstandardized Coefficients B | Std. Error | Standardized Coefficients Beta | t | Sig. | Correlations Zero-order | Partial | Part | Collinearity Statistics Tolerance | VIF |
|---|---|---|---|---|---|---|---|---|---|---|
| 1 (Constant) | 1.020 | .055 | | 18.578 | .000 | | | | | |
| PctCaucasian | −.001 | .000 | −.132 | −7.721 | .000 | −.097 | −.152 | −.127 | .929 | 1.076 |
| mhsupp | .063 | .013 | .102 | 4.690 | .000 | .261 | .093 | .077 | .576 | 1.736 |
| FRLunch | .002 | .000 | .130 | 7.668 | .000 | .090 | .151 | .126 | .947 | 1.056 |
| mhskills | .468 | .018 | .462 | 25.762 | .000 | .495 | .456 | .425 | .845 | 1.183 |
| mhaccess | .068 | .015 | .101 | 4.379 | .000 | .312 | .087 | .072 | .512 | 1.953 |

a. Dependent Variable: mhimpact.

both of these are predictors that are statistically significant, but neither has a large effect size (both have effect sizes = .016).

The results panel shows that the largest contribution to the variance in the impact index is the skills index. Alone, with the other predictors held constant, the skills index accounts for about 18% ($.425^2$) of the variance in the impact index. The other predictors contribute much less to the variance in impact, ranging from .6% ($.077^2$, mhsupport) to 1.6% ($−.127^2$, PctCaucasian). Thus, from this set of data, the combined measure of teachers' technology skills is the best individual predictor of the combined impact on teaching and learning. The results panel also show acceptable levels of "Tolerance" and "VIF" figures for the predictors.

## The "Larger Model" of School Achievement

We have made several references to the nature of our database in school-level achievement, ethnicity, and F/R. As we noted earlier, we used PercentWhite as an index of the proportion of a school's ethnic heterogeneity. We can look at different ethnic proportions (other than PercentWhite) to see how this might affect the outcome in a more complete treatment of the data.

In what follows, I regressed school math achievement on different models of the variables. The first two columns of data in Table 8.9 show the standardized betas and squared part correlations of the nonethnic related variables. The third and fourth columns show the same analysis but adding the ethnic variables, excluding the PercentWhite predictor. Excluding PercentWhite in the second model was necessary to avoid multicollinearity, since the ethnic group proportions of schools are part of the same overall school description, and the four groups make up

**Table 8.9  Regression Output Including and Excluding Ethnic Variables**

| Predictor Variables | Beta[*] | Part[2] | Beta[*] | Part[2] |
|---|---|---|---|---|
| Free or reduced priced meals | −0.60 | 0.19 | −0.54 | 0.13 |
| Students per classroom teacher | −0.07 | 0.00 | −0.06 | 0.00 |
| Avg. years of educational experience | 0.07 | 0.00 | 0.08 | 0.01 |
| Percent teachers with at least master degree | 0.06 | 0.00 | 0.07 | 0.00 |
| Percent white | 0.07 | 0.00 | | |
| PercentAsianPacificIslander | | | 0.14 | 0.01 |
| PercentBlack | | | −0.19 | 0.02 |
| PercentHispanic | | | −0.04 | 0.00 |
| $R^2$ | 0.421 | | 0.444 | |
| (Adjusted) | | | | |
| $F$[a] | 150.97 | | 118.86 | |
| $N$ | 1034 | | 1034 | |

*All standardized beta values were significant at $p < .05$ except PercentHispanic, which was not significant.
[a]F values were significant beyond $p < .001$.

the great majority of the school populations. The Tolerance and VIF figures when PercentWhite was included in the model were .085 and 11.789, respectively. In both cases these figures indicate substantial multicollinearity.

I focused on three ethnic group proportions other than PercentWhite, using the group titles included with the database: "PercentAsianPacific Islander," "PercentBlack," and "PercentHispanic." Again, remember that we cannot interpret the results as individual student findings. Rather, we must interpret these broadly to indicate how different features of the schools in the database relate to achievement results.

The omnibus results for both models are included in the bottom rows of Table 8.9. The findings for both models ($F$ values) were significant, with similar $R^2$ values (from .42 to .44) for both models. Adding the ethnic variables accounted for an additional 2.7% of the variance in math achievement.

There were several noteworthy findings among the individual predictors:

1. The collinearity statistics, without PercentWhite, indicated no meaningful difficulties with the variables in the model, according to our earlier discussion.
2. With the exception of PercentHispanic, each of the predictors was statistically significant.

3. The beta for PercentBlack and PercentHispanic indicated inverse relationships with math achievement when the other predictors were controlled. With PercentHispanic, the relationship was not significant. The significant value for PercentBlack indicated that schools with greater percentages of Black students had lower aggregate math achievement values. However, the effect size was negligible.

4. The squared part correlations indicated that F/R accounted for just over 13% of the variance in math education in model 2 when all other predictors were held constant. This was slightly different from the first model without the additional ethnic variables where the F/R variable accounted for 19% of the variance in the outcome variable.

5. In both models the variables other than F/R did not contribute strongly to the variance in math achievement.

Taken together, these findings are similar to those we discussed above in our multiple regression analyses. The ethnic proportions of schools and other school descriptors contribute differentially to the variance in math achievement of schools with F/R included in the models. Further research is needed that uses student-level information so that we could observe how these and other influences might impact student achievement.

## DISCOVERY LEARNING

In this section we explore how multiple regression is a procedure that builds on bivariate regression by adding additional predictors. We can use multiple regression to show how much of the variance of an outcome variable is accounted for by the set of predictor variables. We can also use it to isolate the impact of individual predictors when the other predictors are held constant. Thus we can observe how much of the variance in the outcome variable is accounted for by a specific predictor when the other predictors are controlled.

We can garner different kinds of information from a multiple regression procedure through different methods of entering the predictors. Entering them in hierarchical fashion according to our theoretical framework allows us to observe the change in the overall $R^2$ as we add each predictor. Some evaluators use a stepwise approach, which allows the statistical program to decide which variables are entered or retained in the analysis. Regardless of the method for entering the predictors, we can use the squared part correlation to understand the unique impact on the "explained variance" of the outcome by specific predictors.

Substantively, our examination of the multiple regression procedure showed similar results to those we found earlier in our discussion of part and partial correlation. In multiple regression results, we found that the ethnic proportions of schools contribute less to the variance in math achievement when we add the proportion of students with low income to the analysis. In the model with two predictor variables, the proportion of students with low incomes accounted for over 16% of the variance in math achievement compared to about 1% accounted for by the ethnic proportion of schools. These differences varied only slightly when ethnicity was defined according to different ethnic group proportions rather than using the generic PercentWhite designation.

## Terms and Concepts

**Hierarchical analyses**   Individual predictors or sets of predictors entered in multiple regression to determine the effects on the $R^2$. The last "model" in the statistical output shows what each predictor (or predictor set) contributes uniquely to the output.

**Multicollinearity**   The extent of the intercorrelation (overlap) among the predictor variables. This is a potentially serious problem for regression methods. When the intercorrelation is high, it makes it difficult to identify what impact each predictor has on the outcome when the others are held constant.

**Order of entry**   Individual predictors entered in multiple regression analyses in a variety of ways. SPSS® indicates "Enter" if the predictors are entered together as a set. Predictors can be added to the model separately by using SPSS® commands (i.e., creating "Blocks" of variables).

**Stepwise regression**   Variables chosen by the statistical program to be entered into and removed from the model according to their impact on the overall probability of the $F$ statistic. This is somewhat controversial to some researchers, since the resulting output could present results different from the output stemming from a theoretically derived sequence of entering predictors. Related procedures (backward elimination and forward entry) also rely on set criteria relating to "probability-of-$F$" values to add or remove variables from the regression model.

**Tolerance** SPSS® statistical procedure for determining the extent of multicollinearity among a set of variables in a regression analysis. Operationally, this measure represents the extent of the intercorrelation of the predictors by subtracting the squared multiple correlation of the additional predictors with a predictor of focus from 1.

**Variance inflation factor (VIF)** A SPSS® statistical calculation that measures the impact of the intercorrelation of predictor variables. If the value is high, it indicates that the intercorrelations have inflated the variance of the regression estimates. Operationally, VIF is (1/ Tolerance).

**Zero-order correlation** The bivariate correlation between two study variables.

## Real World Lab—Multiple Regression

In this exercise, you will use a sample of the 2003 TAGLIT database that I used for the example in this chapter (included in the table below). This time you will regress "mhimpact" (the outcome variable) on two predictors: "mhskills" (the index of teacher computer skills) and "studtch" (a school-level measure of the class size, or student–teacher ratio). Enter the predictors in separate steps so that you can see the change in $R^2$ as each predictor is added to the model.

| studtch | mhskills | mhimpact |
|---------|----------|----------|
| 12.88 | 3.12 | 2.80 |
| 7.38 | 3.12 | 3.05 |
| 15.20 | 3.00 | 2.78 |
| 17.56 | 3.19 | 2.75 |
| 13.60 | 2.68 | 2.46 |
| 11.11 | 3.07 | 2.75 |
| 17.06 | 2.96 | 2.51 |
| 14.77 | 2.45 | 2.31 |
| 13.33 | 2.97 | 2.39 |
| 13.83 | 2.95 | 2.68 |
| 11.20 | 3.19 | 2.88 |
| 9.17 | 2.35 | 2.59 |
| 15.67 | 3.15 | 2.27 |
| 14.59 | 2.85 | 2.66 |
| 12.88 | 3.23 | 2.70 |
| 21.81 | 3.08 | 2.98 |
| 13.53 | 3.18 | 2.46 |

| studtch | mhskills | mhimpact |
|---------|----------|----------|
| 12.34 | 4.00 | 2.88 |
| 12.00 | 3.04 | 2.74 |
| 16.47 | 3.05 | 3.07 |
| 13.33 | 3.26 | 2.51 |
| 13.75 | 3.40 | 2.75 |
| 12.73 | 3.11 | 2.85 |
| 12.12 | 3.40 | 2.75 |
| 17.00 | 3.24 | 2.73 |
| 18.86 | 3.28 | 3.40 |
| 6.69 | 2.83 | 2.49 |
| 27.14 | 3.36 | 2.39 |
| 5.00 | 2.74 | 2.61 |
| 25.00 | 3.20 | 3.41 |
| 13.08 | 3.25 | 2.73 |
| 8.33 | 2.98 | 3.01 |
| 7.36 | 3.43 | 2.86 |
| 8.08 | 3.15 | 3.08 |
| 12.36 | 3.39 | 2.93 |
| 18.43 | 3.18 | 2.43 |
| 15.00 | 3.30 | 3.27 |
| 13.33 | 3.34 | 2.91 |
| 14.44 | 3.23 | 2.62 |
| 18.06 | 3.15 | 2.63 |
| 10.29 | 2.70 | 2.67 |
| 5.10 | 2.62 | 2.89 |
| 9.44 | 2.87 | 2.40 |
| 9.07 | 2.98 | 2.84 |
| 13.89 | 3.23 | 2.86 |

1. Is the overall $F$ value for the model significant? (omnibus test)
2. What is the MLR equation for the final model?
3. How does the $R^2$ value change as new predictors are added?
4. Examine the regression coefficients to determine which are statistically significant. Are the $t$ tests for individual predictors significant?
5. Square the part correlations on the final model when all variables are entered to determine the effect sizes. Which effect size is largest?

# 9

# CODING—USING MULTIPLE REGRESSION WITH CATEGORICAL VARIABLES

Thus far we have discussed multiple regression using continuous variables. But it can also be used in analyses that use categorical variables like gender, ethnic group, class level, or job type. Although multiple regression is not ordinarily used in experimental research, it can certainly be used for analyses that include group membership in experimental or comparative studies. Therefore we need to examine how multiple regression treats categorical variables that typically are coded "1" (referring to membership in a treatment group) or "0" (membership in the comparison group).

## NATURE OF DUMMY VARIABLES

There are several ways to prepare categorical data for MLR analyses, depending on the purpose of an analysis. We will focus here on the most popular of these methods, *dummy coding*. This is a method whereby groups are designated by 1's and 0's so that they can be compared in the multiple regression analysis. Which group is assigned which numeral is arbitrary, but it has a bearing on how you interpret the results. Ordinarily a treatment group is labeled "1" and a control/comparison group labeled "0" in experimental treatments. Whichever group is

*The Program Evaluation Prism: Using Statistical Methods to Discover Patterns*,
by Martin Lee Abbott
Copyright © 2010 John Wiley & Sons, Inc.

labeled "0" becomes the "control" group or "reference" group in the analysis because the study outcome is based on the differences between the category assigned "1" from the reference category.

There are other ways of coding variables according to the purposes of the evaluation research. In particular, "contrast coding" (or "orthogonal coding") is used by the evaluator to specify contrasts by the way in which the variable is coded within the data. Thus, if the evaluator is contrasting the effects of two or more different treatment conditions with a control group, contrast coding would provide a direct method for examining the results for a number of different "pairs" or groups of comparisons among the treatment conditions. "Effect coding" is another method that allows evaluators the opportunity to specify results. For example, the results of effect coding show the differences between a treatment and the "grand mean," whereas dummy coding allows the evaluator to compare the means of the intervention and control groups. Pedhazur (1997) and Cohen et al. (2003), among others, provide thorough explanations of these techniques, including approaches to interpreting the results. I encourage you to explore these authorities as you develop your interests in group comparisons. I offer this exploration of dummy coding to describe the general process of coding and interpretation of coded results.

We discuss regression with only a single, categorical, predictor variable in this section. We will start by examining a single categorical variable that has only two groups so that we can understand the dynamics of interpreting the MLR output. In subsequent sections we will discuss how to create dummy variables when there are more than two groups, and how to interpret such an analysis. Then we will discuss MLR where there are two categorical variables as predictors so that we can examine a more complex interpretation. We should note that using coding schemes with groups of unequal size create different interpretation methods for effect and contrast codes, whereas the outcomes are the same in dummy coding.

## ONE CATEGORICAL VARIABLE WITH TWO GROUPS

MLR treats categorical predictor variables like their continuous counterparts. You can glean additional information from the study output, but the *t* test of the slope coefficient tests the difference between the 1 and 0 groups, much as the groups would be compared in an ANOVA or *t* test. Whatever group is assigned 0 is treated as the comparison or "control" condition.

An example taken from the school achievement dataset we used earlier will show how the process works and what kind of output results from an MLR study. Since our dataset is comprised of aggregated variables, we do not have naturally occurring categorical variables (e.g., gender). Rather, we have variables that identify "percent males" and "percent females" by school. We could use any of several variables for illustration, but I will focus on the "average years of educational experience" of the teachers in the study schools.

Categorizing the educational experience variable can be performed in several ways. One way is to use a *median split* where we create two groups based on the value clusters on either side of the median value. In such a case you would need to obtain the median value in SPSS® through the use of "Analyze–Descriptive Statistics–Frequencies" menus, specifying "median" under the "statistics" tab, shown in Figure 9.1. Note that you can derive this information using two different procedures:

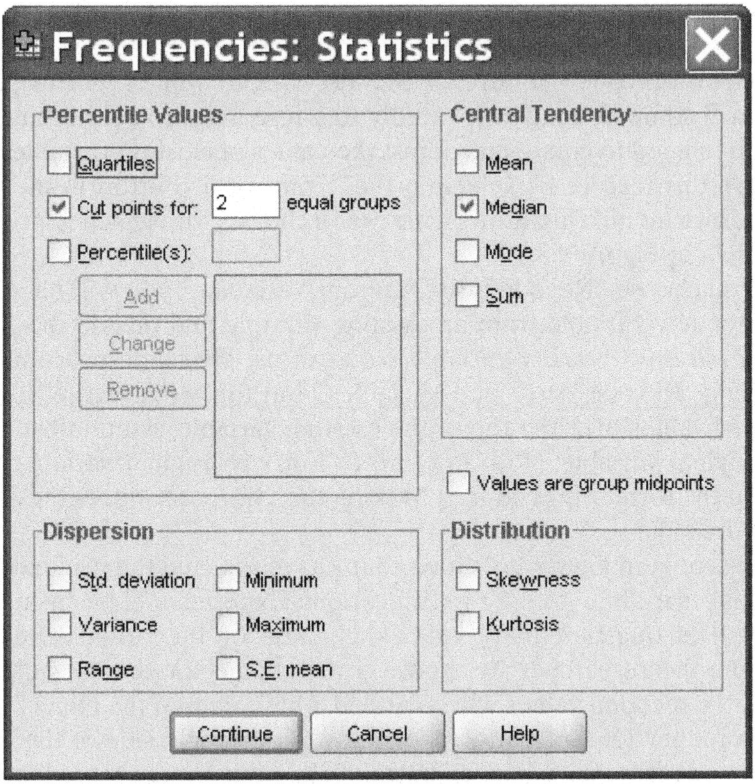

**Figure 9.1**  SPSS® menu for deriving categorizing data.

- You could specify "Median" as shown under "Central Tendency" in Figure 9.1, and then use the median value as the criterion by which to judge whether a particular school's percentage of teachers educational experience is "high" (greater than the median value) or "low" (less than the median value).
- The other method is to select "Cut points for 2 equal groups" as shown in the left panel of Figure 9.1.

Either method allows you to categorize the continuous viable using the median.

Of course, you could use an a priori method to categorize a variable if there is some theoretical rationale or consideration based on past research or experience. Perhaps the educational research literature has suggested that schools with high percentages of teachers with over five years' teaching experience are associated with higher school achievement levels, for example. In this case you would specify five years as the criterion for creating the groups rather than the median.

Having determined the median split value for the variable to be categorized (13.7), you can create a categorical variable by using the "Transform-recode" feature of SPSS®. This extremely useful procedure will come in handy repeatedly in a research project because you will often need to create and adjust the values of existing variables. You can use this feature by selecting the "Transform" button in the main drop-down menu. This allows you several choices with which to proceed, as shown in Figure 9.2.

If you choose "Recode into Different Variables," you will be able to create a new variable from an existing variable and recode the values of the old into the new variable. You can use the same procedure by choosing "Recode into Same Variables," but this would write the new, recoded values over the top of the existing variable, essentially erasing the original variable. (I do not like to lose my original variables, so I choose the former procedure.) Making this choice produces the screen in Figure 9.3.

The screen in Figure 9.3 shows that I have selected the original (continuous) variable "AvgYearsEducationalExperience" by using the arrow. The question mark after the name of the original variable prompts the researcher to choose a new name for the variable into which the recoded values will be added. I have chosen the name "edex" shown in the "Output Variable" window on the right side of the panel. When you push the "Change" button below this window, it will put the new variable name in the window. At that point you may push the

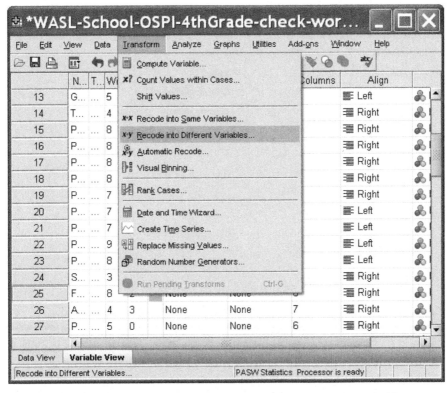

**Figure 9.2**   SPSS® menu for transforming the values of a variable.

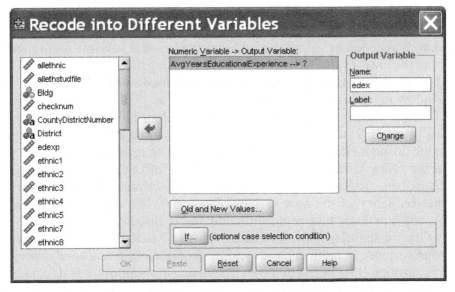

**Figure 9.3**   SPSS® menu for creating a new variable from an existing variable.

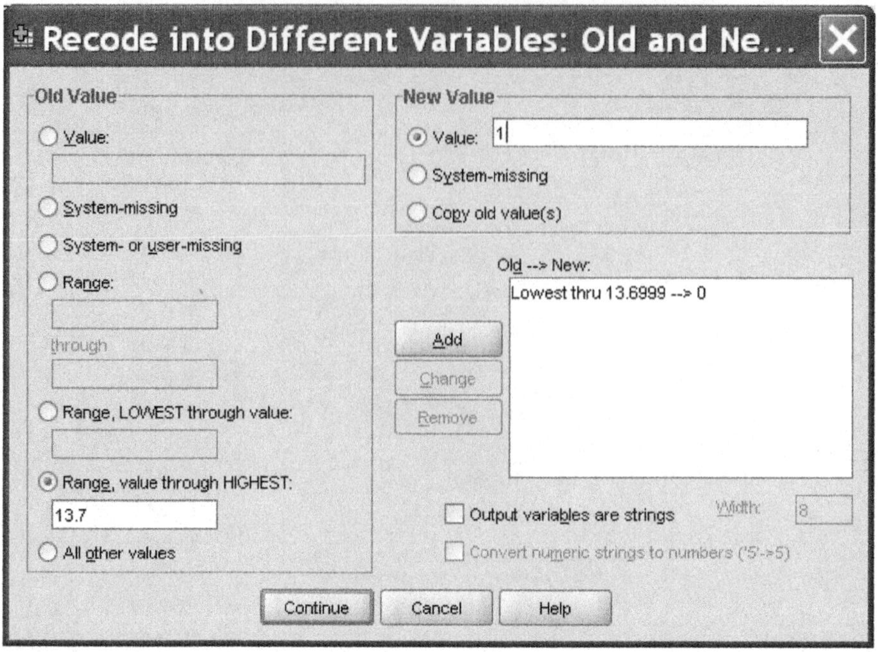

**Figure 9.4** SPSS® menu for recoding the values of an existing variable to a newly created variable.

button "Old and New Values ..." in the bottom center of the window. This will create a new screen as shown in Figure 9.4.

You can use the screen shown in Figure 9.4 to recode the values for the new variable in several ways. In the screen you will see that I have already recoded as "0" all the values in the original variable that range from the lowest value to the value 13.699 by using the "Range, LOWEST through value" button. This recode is described in the "Old→New:" window on the right side of the panel. I chose 13.699 so that I would eliminate the possibility of having original values hitting the categorizing value (13.7) directly. Next, as shown, I specified 13.7 in the "Range, value through HIGHEST:" and specified a "1" to represent those values in the "New Value" window at the top right of the panel. When I hit "Add," the program will record this recode description. Table 9.1 shows the frequency of the new variable categories. The categories are not exactly equal, for they represent comparably distributed values.

Once categorized, the evaluator can use MLR to help determine what variance in math achievement is associated with teachers' experience groups. If we perform the analysis using our school study database, we would get the MLR output in Table 9.2 (using the median value

**Table 9.1    Recoded Values of a Continuous Variable**

| | | | Edexp | | |
|---|---|---|---|---|---|
| | | Frequency | Percent | Valid Percent | Cumulative Percent |
| Valid | .00 | 514 | 49.0 | 49.6 | 49.6 |
| | 1.00 | 522 | 49.7 | 50.4 | 100.0 |
| | Total | 1036 | 98.7 | 100.0 | |
| Missing | System | 14 | 1.3 | | |
| Total | | 1050 | 100.0 | | |

**Table 9.2    MLR Output for One Categorical Predictor**

Model Summary[b]

| | | | | | Change Statistics | | | | |
|---|---|---|---|---|---|---|---|---|---|
| Model | R | R Square | Adjusted R Square | Std. Error of the Estimate | R Square Change | F Change | df1 | df2 | Sig. F Change |
| 1 | .115[a] | .013 | .012 | 16.803 | .013 | 13.779 | 1 | 1034 | .000 |

a. Predictors: (Constant), edexp.
b. Dependent Variable: MathPercentMetStandard.

ANOVA[b]

| Model | | Sum of Squares | df | Mean Square | F | Sig. |
|---|---|---|---|---|---|---|
| 1 | Regression | 3890.560 | 1 | 3890.560 | 13.779 | .000[a] |
| | Residual | 291954.621 | 1034 | 282.355 | | |
| | Total | 295845.181 | 1035 | | | |

a. Predictors: (Constant), edexp.
b. Dependent Variable: MathPercentMetStandard.

Coefficients[a]

| Model | | Unstandardized Coefficients | | Standardized Coefficients | | | Correlations | | |
|---|---|---|---|---|---|---|---|---|---|
| | | B | Std. Error | Beta | t | Sig. | Zero-order | Partial | Part |
| 1 | (Constant) | 56.175 | .741 | | 75.792 | .000 | | | |
| | Edexp | 3.876 | 1.044 | .115 | 3.712 | .000 | .115 | .115 | .115 |

a. Dependent Variable: MathPercentMetStandard.

of 13.7 years of educational experience, labeled "edexp" to categorize the variable).

As shown in the Table 9.2 output, the overall analysis ($F=13.779$, $p < .001$) indicates that the difference between the two groups is

**Table 9.3   Case Summary Output of Categorical Values**

| Case Summaries | | |
|---|---|---|
| MathPercentMetStandard | | |
| edexp | Mean | N |
| .00 | 56.17 | 514 |
| 1.00 | 60.05 | 522 |
| Total | 58.13 | 1036 |

significant, since we only have one predictor in the model. Use the unstandardized B coefficient to see what the difference is between groups "1" and "0". Schools with average educational experience among teachers that is equal to or higher than 13.7 years (those schools classified as "1") are 3.88% higher in school math achievement than the reference category of "0" or schools with less than 13.7 years.

The same information is shown by the $t$ test of edexp in the bottom panel ($t = 3.712$, $p < .001$).[27] Therefore we have determined that the two groups are significantly different, although we can also note that there isn't a great deal of *practical* significance since the overall $R^2$ is only .013.

When we have only the one categorical variable in the analysis, we can use the reported coefficients to understand the group means. Table 9.2 shows that the coefficients are "constant" = 56.175 and "unstandardized B" = 3.876. In this case the constant value is the mean of the "0" group on the outcome variable, while the unstandardized beta represents the difference between the "1" and "0" group means on the outcome variable.

As shown in the SPSS® outcome panel in Table 9.3, the ".00" group has a mean of 56.17, which is equivalent to the constant value noted above, and the difference between the "0" and the "1" group means is 3.88, which is equivalent to the unstandardized beta value (3.876). Therefore schools with educational experience group values higher than 13.7 years are associated with higher school achievement scores than those with less than 13.7 years. There is about a 4 percentage point increase in the achievement results of the "1" group.

The output in Table 9.4 shows that the standard deviation for MathPercentMetStandard is 17.12. Therefore an approximate 4% dif-

---

[27]The squared $t$ value is equivalent to the overall $F$ value ($3.712^2 = 13.779$) in the middle panel, and to the "$F$-Change" value in the top panel.

**Table 9.4   Descriptive Statistics for Math Achievement**

| | | | | | |
|---|---|---|---|---|---|
| | Descriptive Statistics | | | | |
| | N | Minimum | Maximum | Mean | Std. Deviation |
| MathPercentMetStandard | 1050 | 6 | 100 | 57.93 | 17.120 |
| Valid N (listwise) | 1050 | | | | |

ference in math achievement between the schools based on experience is only about one-fourth of a SD; by this criterion, the difference in the groups is not large. Also note from the outcome panels in Table 9.2 that the difference between experience groups, while statistically significant, is not particularly meaningful because it represents only a little more than 1 percent of the variance in math achievement.

It is instructive to compare these results with the outcomes we would obtain if we had decided on a different criterion for categorizing educational experience than the median value of 13.7. Try different cutoff criteria with your own data set to see how the outcomes of the analysis will differ depending on the categorizing criteria. *"Splitting" the continuous variable usually results in different outcomes depending on the way in which the predictor variable was categorized.*

## CREATING DUMMY VARIABLES

Creating dummy variables is not difficult if you remember that you are going to create separate *subvariables* (variables within a variable) for each group of the categorical variable in question. For example, if you have a categorical variable with three groups, you create a column of data corresponding to each of the groups. Each of these separate parts of the variable is a subvariable. You have to remember that the overall variable is comprised of all the subvariables. It is perhaps easier to show how the data file might look for a categorical variable comprised of three groups.

Table 9.5 shows there is one outcome variable ("outcome"), and one categorical variable ("treatment"). There are also three subvariables, one for each of the groups of the categorical variable. Looking at the first subvariable (dummy subvariable one, or "dsubvar1"), you can see that the cases or case values are coded as "1" when the categorical variable treatment is "1." The remaining cases on that subvariable are "0." In essence, we are creating a subvariable that is "identified" as group 1. The additional two subvariables are coded "1" corresponding

**Table 9.5   Example Data for Creating Categorical "Subvariables"**

| treatment | outcome | dsubvar1 | dsubvar2 | dsubvar3 |
|-----------|---------|----------|----------|----------|
| 1 | 6 | 1 | 0 | 0 |
| 1 | 7 | 1 | 0 | 0 |
| 1 | 5 | 1 | 0 | 0 |
| 1 | 6 | 1 | 0 | 0 |
| 1 | 7 | 1 | 0 | 0 |
| 2 | 9 | 0 | 1 | 0 |
| 2 | 10 | 0 | 1 | 0 |
| 2 | 8 | 0 | 1 | 0 |
| 2 | 9 | 0 | 1 | 0 |
| 2 | 10 | 0 | 1 | 0 |
| 3 | 3 | 0 | 0 | 1 |
| 3 | 4 | 0 | 0 | 1 |
| 3 | 2 | 0 | 0 | 1 |
| 3 | 3 | 0 | 0 | 1 |
| 3 | 4 | 0 | 0 | 1 |

to the group number on the categorical variable treatment. Thus, for example, dsubvar2 shows "1's" only when the categorical variable treatment shows "2." The same is true for the third subvariable.

What we have done is to create three variables within a variable (subvariables) from one categorical variable with three groups. Each of the subvariables represents one of the groups, or identify the groups. Actually only two subvariables are needed to represent this variable, since if you know the values of two, you know the values of the third without identifying it. In fact, when you run the MLR analysis, you typically do not enter the last variable. It becomes the "0" reference category against which the other groups are compared. This will be easier to see when we take an example. The only consideration to keep in mind when you do the MLR analysis is that the subvariables together make up the variable; they are not separate variables.

## CREATING SUBVARIABLES IN SPSS®

It is easy to create subvariables with SPSS®. To see this, we will create a string of operations that may be helpful using the data shown in Table 9.5 without the subvariables, as shown in Table 9.6.

We start by creating three variables (which will become the subvariables) and make all the values equal to 0. You can do this in SPSS® by

**Table 9.6   Categorical Data without Dummy Categories**

| treatment | Outcome |
|-----------|---------|
| 1 | 6 |
| 1 | 7 |
| 1 | 5 |
| 1 | 6 |
| 1 | 7 |
| 2 | 9 |
| 2 | 10 |
| 2 | 8 |
| 2 | 9 |
| 2 | 10 |
| 3 | 3 |
| 3 | 4 |
| 3 | 2 |
| 3 | 3 |
| 3 | 4 |

using the "Transform" menu shown in Figure 9.2. Selecting "Compute" will create the screen in Figure 9.5. Using this procedure, you can specify a new variable to be created in the "Target Variable:" window. In Figure 9.5, I show that I am creating a variable (subvariable) called "dsubvar1" and that it has a value of "0." I would repeat this process three times to create all three subvariables ("dsubvar1," "dsubvar2," and "dsubvar3"). If you are familiar with SPSS® syntax, the following will accomplish the same results:

```
compute dsubvar1 = 0.
compute dsubvar2 = 0.
compute dsubvar3 = 0.
execute.
```

Once all the subvariables are created, the data file will look like the file in Table 9.6a. At this stage all the subvariable values are "0," and you can proceed to assign values to them based on the values of the treatment (categorical) variable. This will complete the process of creating dummy categories (subvariables) from the treatment variable.

The next procedure is to give the cases on dsubvar1 a value of "1" when values of the variable "treatment" are "1." In the same way, give cases on dsubvar2 and dsubvar3 of "1" when values of treatment are 2 or 3, respectively. This process is easy, but it takes a few steps. First,

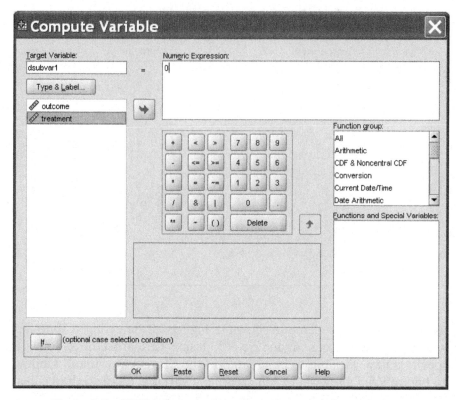

**Figure 9.5** SPSS® "compute" procedure for creating subvariables.

**Table 9.6a Data File with Variables and Subvariables**

| treatment | outcome | dsubvar1 | dsubvar2 | dsubvar3 |
|---|---|---|---|---|
| 1 | 6 | 0 | 0 | 0 |
| 1 | 7 | 0 | 0 | 0 |
| 1 | 5 | 0 | 0 | 0 |
| 1 | 6 | 0 | 0 | 0 |
| 1 | 7 | 0 | 0 | 0 |
| 2 | 9 | 0 | 0 | 0 |
| 2 | 10 | 0 | 0 | 0 |
| 2 | 8 | 0 | 0 | 0 |
| 2 | 9 | 0 | 0 | 0 |
| 2 | 10 | 0 | 0 | 0 |
| 3 | 3 | 0 | 0 | 0 |
| 3 | 4 | 0 | 0 | 0 |
| 3 | 2 | 0 | 0 | 0 |
| 3 | 3 | 0 | 0 | 0 |
| 3 | 4 | 0 | 0 | 0 |

use the "Transform–Recode Into Same Variables" that we discussed above. This time we are simply "replacing" "0's" in the dsubvar1 variable with "1's" when the treatment variable is 1.

The SPSS® panel shown in Figure 9.6a indicates that we are identifying "dsubvar1" to be recoded, and that we will recode it conditionally on the value of the treatment variable. This latter part of the process is created by using the "If ... optional case selection condition" button in the lower part of the panel. Choosing this button yields the information included in the screen in Figure 9.6b.

As Figure 9.6b shows, we are requesting a recode of the dsubvar1 value "0" to another value when treatment=1. Hitting "Continue" at the bottom of this screen allows us to use the "Old and New Values" button shown in the lower part of the screen of Figure 9.6a. This part of the process will then be the same process we used above in Figure 9.4 to recode the values of an existing variable. In the present case we will create the screen shown in Figure 9.6c. In this example I have indicated that the "0" values of dsubvar1 should now be coded "1" when treatment=1. The other "0" values would remain "0."

Using the same procedure on the other subvariables would yield the database shown in Table 9.7. The database would include the original treatment and outcome variables along with the three subvariables, which would be coded with values of "1" corresponding to the treatment groups of "1," "2," and "3". The same procedure holds true for categorical variables with more than three groups. Simply create the

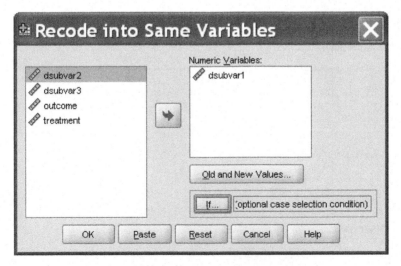

**Figure 9.6a** Recoding an existing variable.

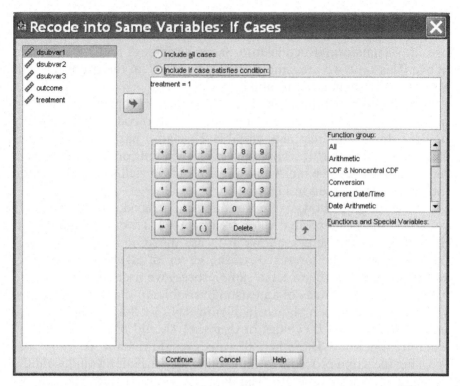

**Figure 9.6b** Recoding an existing variable using conditional selection.

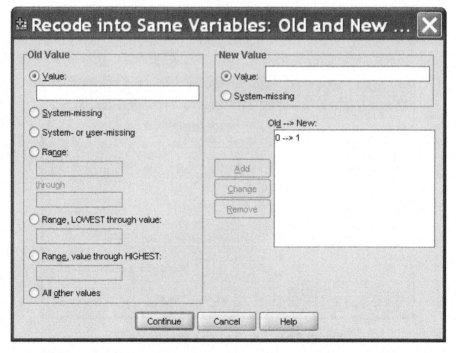

**Figure 9.6c** Recoding conditional values to the existing (sub)variable.

**Table 9.7   Final Database Showing Recoded Subvariables**

| treatment | outcome | dsubvar1 | dsubvar2 | dsubvar3 |
|---|---|---|---|---|
| 1 | 6 | 1 | 0 | 0 |
| 1 | 7 | 1 | 0 | 0 |
| 1 | 5 | 1 | 0 | 0 |
| 1 | 6 | 1 | 0 | 0 |
| 1 | 7 | 1 | 0 | 0 |
| 2 | 9 | 0 | 1 | 0 |
| 2 | 10 | 0 | 1 | 0 |
| 2 | 8 | 0 | 1 | 0 |
| 2 | 9 | 0 | 1 | 0 |
| 2 | 10 | 0 | 1 | 0 |
| 3 | 3 | 0 | 0 | 1 |
| 3 | 4 | 0 | 0 | 1 |
| 3 | 2 | 0 | 0 | 1 |
| 3 | 3 | 0 | 0 | 1 |
| 3 | 4 | 0 | 0 | 1 |

corresponding subvariables identifying the groups by the use of "1" and "0" values.

Again, if you are familiar with SPSS® syntax, you can use the following to accomplish the same result:

```
if (treatment = 1) dsubvar1 = 1.
if (treatment = 2) dsubvar2 = 1.
if (treatment = 3) dsubvar3 = 1.
execute.
```

We are now ready to proceed with the MLR analysis. In the next section we show how this works using only the one variable (composed of three subvariables) as a predictor.

## ONE CATEGORICAL VARIABLE WITH MORE THAN TWO GROUPS

If a categorical variable has more than two categories or groups, the method for interpreting the SPSS® results is very similar for analyses with two categories. The main thing to remember is that you enter all but one of the subvariables into the analysis. In the example of the previous section, we created three subvariables from the treatment variable. In the MLR analysis, we would only enter two of these subvariables.

If we were to enter all three subvariables, we would create an untenable condition of multicollinearity. In this condition the computer program would interpret the subvariables as separate predictor variables, which would be so highly intercorrelated (since they are part of the same variable) that the resulting regression coefficients would not be reliable. As we noted in Chapter 8, multicollinearity generally is a problem when the predictor variables are highly correlated with each other. But when all the subvariables are in the same analysis, the intercorrelations cause problems. In the latest SPSS® versions, one of the subvariables is dropped from the analysis if all three are inadvertently included, so be sure to examine the output carefully.

Using our example, we would enter two subvariables, the choice of which depends on the nature of the research problem. *Whichever subvariable is not entered becomes the reference category the other subvariables are compared against.* The following are features of the output:

- The intercept (or "constant") term in the MLR output is the mean value on the outcome variable of the subvariable not in the equation.
- Each of the "B" coefficients represents the difference in means between the subvariable in question from the mean of the subvariable not in the equation.
- The overall $R^2$ is the same, no matter which groups are "identified" in the subvariables, and is a test of the assumption that each of the group means is equal.
- Testing the significance of each b is essentially testing the difference between the means of the identified subvariable and the subvariable not in the equation on the outcome variable.

## A Hypothetical Example

The SPSS® output of an MLR analysis of the example data is shown in Table 9.8. In this case we are predicting the value of the "outcome" variable using two of the three subvariables of the predictor variable "treatment." For illustration purposes we have used "dsubvar1" and "dsubvar2." Remember, "dsubvar3" is not included in the analysis and becomes the reference group against which the other two subvariables are compared.

1. The "constant" estimate of 3.2 represents the mean value on the outcome variable of the subvariable not in the equation. You can

**Table 9.8    MLR Output for Categorical Subvariables**

Model Summary

| | | | | | | Change Statistics | | | |
|---|---|---|---|---|---|---|---|---|---|
| Model | R | R Square | Adjusted R Square | Std. Error of the Estimate | R Square Change | F Change | df1 | df2 | Sig. F Change |
| 1 | .956[a] | .915 | .900 | .837 | .915 | 64.286 | 2 | 12 | .000 |

a. Predictors: (Constant), dsubvar2, dsubvar1.

Coefficients[a]

| | | Unstandardized Coefficients | | Standardized Coefficients | | | Correlations | | |
|---|---|---|---|---|---|---|---|---|---|
| Model | | B | Std. Error | Beta | t | Sig. | Zero-order | Partial | Part |
| 1 | (Constant) | 3.200 | .374 | | 8.552 | .000 | | | |
| | dsubvar1 | 3.000 | .529 | .552 | 5.669 | .000 | .000 | .853 | .478 |
| | dsubvar2 | 6.000 | .529 | 1.104 | 11.339 | .000 | .828 | .956 | .956 |

a. Dependent Variable: outcome.

use Table 9.7 to see how the third subvariable group mean is created:

$$(3+4+2+3+4)/5 = 3.2.$$

2. Each of the unstandardized B coefficients represents the mean differences between the subvariables identified and the subvariable not in the equation:

$$B(\text{dsubvar1}) = 3.0\,(6.2 - 3.2)$$
$$B(\text{dsubvar2}) = 6.0\,(9.2 - 3.2)$$

3. The corresponding $t$ values indicate that both of these subvariables are significantly different from the subvariable not in the equation.

4. The $R^2$ of .915 ($F = 64.286, p < .001$) is statistically significant, indicating that the group means of the subvariables are not equal, and the group differences account for .915 of the variance in the outcome (not surprising given the small $N$ size and nature of the example database).

## An Example from the School Database

In the preceding example with educational experience and school-level achievement, we used a median split process to arrive at two groups of values of the average educational experience by school. To illustrate how the process would work with a categorical variable with three (or more) groups, I created three groups from the educational experience variable using the SPSS® method of creating equal groups in the "Analyze–Descriptive Statistics–Frequencies–Statistics" menu.

This resulted in a variable "edexperience" that has three subvariables, shown in Table 9.9 along with the mean values of each on the outcome variable school math achievement. Table 9.10 shows the MLR analysis with subvariables one and two ("edexp1" and "edexp2"). I will only show the estimated coefficients to illustrate the elements we discussed earlier.

If you refer to the school math achievement means in Table 9.9, you will find that the unstandardized B coefficients represent the differences between the edexperience groups (subvariables) according to the process we discussed earlier (accounting for rounding differences). Remember, the third subvariable ("edexp3") is not shown in the analysis and is therefore the reference group:

**Table 9.9** **Descriptive Output for Years of Educational Experience Subvariables**

| Subvariable | Average Years of Educational Experience by School | Mean Values of Subvariables on "%MetMathStandard" |
| --- | --- | --- |
| edexp1 | Up to 12.5 years | 55.15 |
| edexp2 | 12.5 to 15 years | 59.15 |
| edexp3 | 15 years and over | 60.06 |

**Table 9.10** **MLR Output for Educational Experience Subvariables**

| | | Coefficients[a] | | | | | | | |
| --- | --- | --- | --- | --- | --- | --- | --- | --- | --- |
| | | Unstandardized Coefficients | | Standardized Coefficients | | | Correlations | | |
| Model | | B | Std. Error | Beta | t | Sig. | Zero-order | Partial | Part |
| 1 | (Constant) | 60.063 | .897 | | 66.936 | .000 | | | |
| | Edexp1 | −4.917 | 1.275 | −.137 | −3.858 | .000 | −.124 | −.119 | −.119 |
| | Edexp2 | −.918 | 1.276 | −.026 | −.720 | .472 | .042 | −.022 | −.022 |

a. Dependent Variable: MathPercentMetStandard.

$$\text{Constant} = 60.063 \, (\text{the mean for edexp3 not in the analysis})$$
$$B(\text{edexp1}) = (-4.917) = (55.15 - 60.06)$$
$$B(\text{edexp2}) = (-.918) = (59.15 - 60.06)$$

## DISCOVERY LEARNING

Continuous variables can be coded into two or more categorical groups for certain analyses, which is like a comparison of treatment groups. Dummy coding is a way to code variables into subvariables so that the categorical groups can be compared.

In an attempt to be more "accurate," researchers often assume that creating more groups from a continuous variable might capture more of the information contained in the continuous variable that is categorized. Thus some might create "high," "medium," and "low" categories instead of just high and low. Regardless of the number of such groups the process could be considered artifactual, depending on the research problem, so it can be misleading. At the very least, valuable information may be lost as a result.

In the last section we used the example of creating groups from the continuous variable of years of educational experience to demonstrate MLR with categorical data. This example was only for demonstration using real data, and the results were instructive. When we used a median split (resulting in two groups), the identified group ("1," or "high") was 3.88% higher on math achievement than the nonidentified group ("0," or "low"). When we created three groups instead of two, we found that the difference between the lowest ("1") and highest ("3") groups of average school-level educational experience was approximately 5%. Neither finding was particularly meaningful, as shown by the effect sizes, but you can see how creating different groups yields different results.

## DETAIL FOR THE CURIOUS—FALSE DICHOTOMIES

Before we look at an example, it is important to discuss *false dichotomies* or categorical variables that have been created by "dichotomizing" continuous variables. Some authorities do not favor creating categorical variables from continuous variables because these methods do not allow the researcher to "use" all of the data. Evaluation researchers, however, are often faced with situations where it may be useful to do so, for example, when they graphically present data in groups. We leave it to the evaluator to assess the situations where it is appropriate to use these methods.

Creating categories (e.g., "large" and "small") that do not occur in this fashion in the raw data is one example of this process. It is popular among some evaluators to perform a "median split" or some other procedure whereby they create two or more categories out of a continuous variable in order to use it in the analysis as "high" versus "low" or related categories.

To discuss this process with variables available in the achievement dataset example, we might be interested in comparing the impact of schools with "high" and "low" enrollments on achievement. In this case an evaluator would split the "total enrollment" variable in the school outcome database in order to compare "small" and "large" schools in an ANOVA procedure. Using this process would not capture all the information contained in the continuous variable that might be helpful in the overall analysis. Categorizing total enrollment into two groups would mask the variation within each of the resulting categories. So, rather than having 1000 data points (as you would have in a study with a study size of 1000), you would have two.

There may be patterns in the variation of the raw data that cannot be adequately captured with two categories. This might create results that would not be as powerful as they could be. The nature of the results would then be due to the method for categorizing rather than due to the naturally occurring variation in the data.

## Terms and Concepts

**Contrast coding**   Group codes assigned such that specific contrasts between groups of interest are built in to the coding scheme. The groups are identified by the assignment of 1's and 0's as in the dummy and effect coding, but depending on the contrast of interest, the groups can be assigned other "weights" like −1 and −2. The weights for each subvariable equal "0" when summed.

**Dummy coding**   A categorical variable for coding into dichotomous groups identified by 1's and 0's, where a category of interest, or "treatment" is coded a "1" and the additional group(s) coded "0." The group coded all "0's" becomes the comparison group.

**Effect coding**   A coding scheme similar to dummy coding, except the subvariables contain a (last) group coded "(–1)" in addition to the other groups coded "1's" and "0's". This enables the evaluator the opportunity to specify results. For example, the results of effect coding show the differences between a treatment and the "grand mean," whereas dummy coding allows the evaluator to compare the means of the intervention and control groups.

**False dichotomies**   Categorical variables that have been created by "dichotomizing" continuous variables. Some authorities do not favor creating categorical variables from continuous variables because these methods do not allow the researcher to "use" all of the data from the continuous variable.

**Median split**   A method to create a dichotomous variable from a continuous variable by using the median of the continuous variable to define "high" and "low" values for use in subsequent statistical analyses.

**Subvariables created in dummy coding**   "Variables within a variable" for each group created in dummy coding. The variable to be coded is used to create several subvariables depending on the number of groups. The number of subvariables created is the number of groups in the variable in question minus 1. If the evaluator has a "treatment" variable identifying three values (e.g., comparing "traditional" with "ethnomathematics" and "constructivist" math teaching approaches), there would be two dummy subvariables created and used in the MLR analysis.

## Real World Lab—Dummy Coding

In this exercise you will use a database to create a dummy variable with three groups and then use the resulting data in a regression equation. The data for the exercise in the following table are adapted from an educational database in which have been changed to represent three "treatments" used to predict a group outcome. For this example we will assume that the three treatment groups are different teaching approaches and the outcome is a measure of student math achievement. The variables are "Trtmtgroups" representing three methods of teaching math: 1=traditional approach, 2=ethnomath approach, 3=constructivist approach. "Classrating" an individual student achievement assessment in math with possible scores ranging from between 0 and 150.

| classrating | trtmtgroups |
|:-----------:|:-----------:|
| 18  | 1 |
| 40  | 1 |
| 42  | 1 |
| 15  | 2 |
| 22  | 2 |
| 30  | 2 |
| 32  | 2 |
| 27  | 3 |
| 45  | 1 |
| 17  | 2 |
| 50  | 3 |
| 21  | 2 |
| 58  | 3 |
| 5   | 2 |
| 31  | 2 |
| 26  | 3 |
| 120 | 3 |
| 14  | 2 |
| 50  | 1 |
| 34  | 3 |
| 102 | 3 |
| 105 | 3 |
| 63  | 3 |
| 64  | 1 |
| 12  | 1 |
| 50  | 1 |

Create subvariables for the trtmtgroups variable and use them in a regression equation predicting classrating:

1. What are the omnibus and individual predictor results?
2. How do you interpret the constant and unstandardized B values?

# 10

# INTERACTION

Evaluators who use multiple regression often run into situations where the regression of an outcome variable on a predictor variable is differentially affected by another predictor variable. That is, when the relationship between one predictor and the outcome variable changes at different levels of another predictor variable. This is an "interaction" effect.

As an example suppose that we found that there was a significant relationship between attitudes toward the workplace (predictor) and job satisfaction (outcome). An interaction effect is present if the high and low values on a separate predictor variable (salary) significantly affected the nature of the original regression equation. As shown in Figure 10.1, highly paid workers show a much stronger positive relationship between their company attitudes and work satisfaction than those not so highly paid. This kind of relationship among the study variables may not be immediately apparent.

In this chapter I use two examples of interactions: one example shows the interaction in a MLR analysis with two continuous predictors, and the other example shows an interaction in a MLR with a continuous and a categorical predictor. You can identify interactions using MLR, but it is often easiest to see when interactions are present

*The Program Evaluation Prism: Using Statistical Methods to Discover Patterns,*
by Martin Lee Abbott
Copyright © 2010 John Wiley & Sons, Inc.

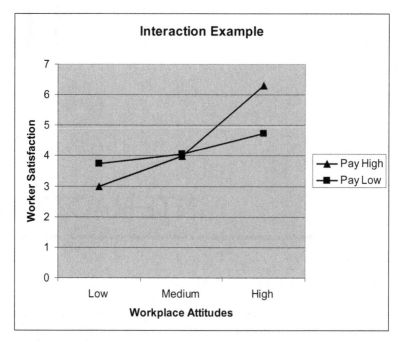

**Figure 10.1**  Interaction example between workplace attitudes and worker pay predicting worker satisfaction.

by using graphical methods. I will show how to create graphs from the data so that you can see the interactions.

Understanding and analyzing the nuances of interactions is complex. In this chapter I cover the basics, but you should consider further study if you are interested in all the dimensions of the process. I note, in particular, three resources for studying interactions that are very thorough: Aiken and West (1991), Pedhazur (1997), and Cohen et al. (2003). These and other resources on the topic demonstrate how to calculate the point of intersection between different group regression lines, and the nature of the interaction, among other things.

Interactions can be "ordinal" or "disordinal," for example, depending on the trends of the separate regression lines. Ordinal interactions show separate group regression lines that are not parallel, but may not actually meet or "cross" within the range of the data used to create the analysis. Disordinal interactions show group regression lines that cross within the range of values used to create them. Group regression lines that are parallel do not represent an interaction condition.

Detecting and interpreting interactions using MLR is sometimes difficult and depends a good deal on the nature of the data. Evaluators interested in a comprehensive treatment of interaction might seek advanced statistical works that focus specifically on the details of such relationships. In this book we only discuss some of the main approaches to using MLR with interactions.

## INTERACTIONS WITH CONTINUOUS VARIABLES

In order to analyze interactions, it is first important to recognize whether or not they are actually present! If you suspect an interaction due to a priori reasons or informed suspicions, you might check the possibilities by creating *product terms* for the appropriate variables and running an MLR analysis. The product term is simply a variable that consists of multiplying the values of the variables that you suspect may interact.

We will distinguish between product terms involving continuous and categorical data later, but for continuous variables, run the MLR with the product term to see if it is a significant predictor of the outcome. If so, you will need to change your approach to the analysis. Product terms also exist for more than two variables, but the interpretation for more than two or three "levels" is very difficult. In what follows, we will limit our discussion to a single product term involving two variables.

When assessing an MLR relationship with continuous variables, it is often helpful first to "center" the predictors. *Centering* refers to subtracting a constant amount from each variable, typically using the mean of each variable. The result is a set of values that, when summed, will equal zero, or near zero (when using mean centering). The advantage of centering is that you will reduce the possibility of multicollinearity among the predictors when there is an interaction, thereby allowing you to have more confidence in the resulting regression estimates.[28] Stated differently, this will reduce the correlations between the "product term" and the predictor variables so as to reduce multicollinearity. The standard deviations of the centered variables will be unchanged, but the regression estimates will change.

Here are the main steps in conducting a regression analysis where there is an interaction present with continuous variables:

[28]The authorities on using centering to identify and control multicollinearity are Aiken and West (1991), and Cohen et al. (2003). Both sources provide a thorough treatment of the method and effects of using centering in interactions. I strongly encourage readers to review these sources for a definitive treatment of the subject.

**Table 10.1  Five Cases of Database with Study Variables**

| gensat | desiremgmt | attmgmt | desmgcen | attmgrcen | xzcen |
|--------|-----------|---------|----------|-----------|-------|
| 5.80 | 4.00 | 5.00 | 1.64 | 0.87 | 1.43 |
| 3.40 | 4.00 | 4.33 | 1.64 | 0.20 | 0.33 |
| 3.60 | 3.00 | 4.00 | 0.64 | −0.13 | −0.08 |
| 2.80 | 1.33 | 3.00 | −1.03 | −1.13 | 1.16 |
| 2.80 | 1.00 | 2.67 | −1.36 | −1.46 | 1.99 |

1. Center the predictor variables in question by subtracting the mean of each variable from its values.
2. Create a *product term* that consists of multiplying the centered values of the predictor variables in question.
3. Perform an MLR analysis in order to detect whether or not the interaction term (i.e., product term) is significant.
4. If the product term is significant, substitute values for one of the variables and create separate regression equations so that you can observe the interaction graphically.

We will examine these steps by using a database on worker participation that I collected in conjunction with an earlier evaluation study. Table 10.1 includes five of the cases and shows the study variables as raw data values and centered values.

To take an example, the first case shows that the value for the variable "attmgmt" (attitudes toward management[29]) is 5.00. When the mean of that variable (4.129; taken from the entire dataset) is subtracted from this case, the result is a new variable "attmgrcen" (centered values of attitudes toward management) with a value on the first case of 0.87 (5.00–4.129). Similarly the 4.00 value for the first case of "desiremgmt" (desire for participation in decision making[30]) becomes (1.64) on the new variable "descen" when it is centered by subtracting its mean value (2.3627; taken from the entire database). Multiplying these two centered values together results in a centered product term, "xzcen."

[29]"Attmgmt" is an index created from a principal components analysis that combines three items. Higher scores (on a scale ranging from "1–Strongly Disagree" to "6–Strongly Agree") indicate greater support for the management role.
[30]"Desire" is a factor resulting from a principal components analysis that combines three items (on a scale ranging from 1–"None" to 4–"A lot"), with higher scores indicating stronger desire for making decisions about worker promotion, firing, and handling complaints.

**Table 10.2  Correlation Matrix with Centered and Uncentered Variables**

|            | gensat  | attmgmt | desiremgmt | xzuncent | attmgrcen | desmgcen | xzcen  |
|------------|---------|---------|------------|----------|-----------|----------|--------|
| gensat     | 1.000   | .330**  | .035       | .211*    | .330**    | .035     | .224*  |
| attmgmt    | .330**  | 1.000   | .004       | .484**   | 1.000**   | .004     | −.005  |
| desiremgmt | .035    | .004    | 1.000      | .860**   | .004      | 1.000**  | .037   |
| xzuncent   | .211*   | .484**  | .860**     | 1.000    | .484**    | .860**   | .201*  |
| attmgrcen  | .330**  | 1.000** | .004       | .484**   | 1.000     | .004     | −.005  |
| desmgcen   | .035    | .004    | 1.000**    | .860**   | .004      | 1.000    | .037   |
| xzcen      | .224*   | −.005   | .037       | .201*    | −.005     | .037     | 1.000  |

Table 10.2 shows the correlations among the centered and uncentered variables. In particular, note the correlations highlighted in the fourth row of data (.484 and .86), which are the zero-order correlations between the uncentered product term and, respectively, the uncentered variables "attmgmt" and "desiremgmt." When these variables are centered, the correlations change dramatically, as shown in the bottom row of data (−.005 and .037). As noted above, centering helps reduce multicollinearity and ensures more confidence in the regression estimates. Also note in the table that the correlations between both of these variables with "gensat" (job satisfaction) do not change when the variables are centered, as shown in the shaded cells of the first column of data.

The MLR analysis with the centered variables predicting job satisfaction is shown in Table 10.3. The results indicate that the product term is significant (Sig. $F$ Change = .038, $t$ = 2.105), which suggests that there is an interaction between the two predictor variables "attmgrcen" and "desmgcen" in their prediction of "gensat." Technically the results for model 2, showing only the two predictors without the product term, identifies the "main effects." Some statistical authorities look at these separately before looking for a possible interaction. However, a significant interaction makes the interpretation of the predictor variables more complex, since values of one predictor and the outcome variable changes at different levels of another predictor variable. I therefore look for a significant interaction before I interpret the main effects.

The resulting regression equation is

$$\text{Gensat} = (.411)_{\text{attmgrcen}} + .031_{\text{desmgcen}} + .306_{\text{xzcen}} + 4.047$$

When the interaction term is significant, it is helpful to substitute values of one of the predictors and create separate regression equations so that you can graph the results. You can do this through directly

**Table 10.3   MLR Output with Centered Variables Predicting Job Satisfaction**

Model Summary

| Model | R | R Square | Adjusted R Square | Std. Error of the Estimate | Change Statistics | | | | |
|-------|---|----------|-------------------|----------------------------|------------------|---|-----|-----|--------------|
| | | | | | R Square Change | F Change | df1 | df2 | Sig. F Change |
| 1 | .323[a] | .104 | .096 | 1.20538 | .104 | 11.990 | 1 | 103 | .001 |
| 2 | .324[b] | .105 | .088 | 1.21063 | .001 | .109 | 1 | 102 | .742 |
| 3 | .378[c] | .143 | .117 | 1.19077 | .038 | 4.432 | 1 | 101 | .038 |

a. Predictors: (Constant), attmgrcen.
b. Predictors: (Constant), attmgrcen, desmgcen.
c. Predictors: (Constant), attmgrcen, desmgcen, xzcen.

Coefficients[a]

| Model | | Unstandardized Coefficients | | Standardized Coefficients | t | Sig. | Correlations | | |
|-------|---|-----------------------------|-----------|---------------------------|---|------|--------------|---|---|
| | | B | Std. Error | Beta | | | Zero-order | Partial | Part |
| 1 | (Constant) | 4.045 | .118 | | 34.357 | .000 | | | |
| | attmgrcen | .435 | .126 | .323 | 3.463 | .001 | .323 | .323 | .323 |
| 2 | (Constant) | 4.044 | .118 | | 34.199 | .000 | | | |
| | attmgrcen | .435 | .126 | .323 | 3.451 | .001 | .323 | .323 | .323 |
| | desmgcen | .039 | .119 | .031 | .330 | .742 | .027 | .033 | .031 |
| 3 | (Constant) | 4.047 | .116 | | 34.792 | .000 | | | |
| | attmgrcen | .411 | .125 | .305 | 3.295 | .001 | .323 | .312 | .304 |
| | desmgcen | .031 | .117 | .024 | .260 | .795 | .027 | .026 | .024 |
| | xzcen | .306 | .145 | .195 | 2.105 | .038 | .224 | .205 | .194 |

a. Dependent Variable: gensat.

substituting values into the equation above, but Aiken and West (1991) describe a process for doing this through SPSS®. For example, if we want to see how high and low values of "desmgcen" affect the relationship between "attmgrcen" and "gensat," with this process we can choose one standard deviation (SD) of "desmgcen" to use for calculating high and low values in the regression equation.

We generate high and low values by subtracting the SD when it is high (+SD) and when it is low (−SD). First, we would subtract the (+SD) from values of "descen" to form a new variable called "descenzh" to indicate that this value represents a "high" value of Z. This represents "high" since we are using the positive value of the SD as the number that drives the result. The following is an example using a value of desmgcen from the database:

$$\text{desmgcen}\,(-.70) - (+.987) = -1.687$$

We would then need to create another product term with "att-mgrcen," since we would be changing the original variable used to create the product term. This would result in the new product term, which we can call "xzhicen" for example, since it is the product term for the two predictors $X$ and $Z$.

Next, regressing "gensat" with "attmgrcen" (the original centered predictor), "descenzh" (the high value of "desmgcen"), and the new product term "xzhicen" would yield a new regression equation. We would only be concerned with the new constant and the regression estimate for the unchanged predictor ("attmgrcen). Aiken and West (1991) refer to these as "simple slopes" or the regression line slope between predictor and outcome at a specific value of a second predictor (other authorities refer to these as "simple effects"). In this case our regression results where "desmgcen" is a *high value* is shown in Table 10.4:

$$\text{Gensat} = (.713)_{\text{attmgrcen}} + 4.077$$

If we follow the same procedure to create a new low value for "desmgcen" by subtracting the $(-1\,\text{SD})$ of "desmgcen" from values of "desmgcen," we would create a new variable called "descenzl," to indicate *low values* of this predictor. The following is the same data point we used above to illustrate the calculation of the high value. In this case, we are subtracting the low $(-\text{SD})$ value:

$$\text{desmgcen}\,(-.70) - (-.987) = .287$$

**Table 10.4  MLR Output for Simple Slope of attmgrcen with Job Satisfaction When desmgcen Is High**

| | | | | | | | | |
|---|---|---|---|---|---|---|---|---|
| | Coefficients[a] | | | | | | | |
| | Unstandardized Coefficients | | Standardized Coefficients | | | Correlations | | |
| Model | B | Std. Error | Beta | t | Sig. | Zero-order | Partial | Part |
| 1 (Constant) | 4.077 | .163 | | 24.991 | .000 | | | |
| attmgrcen | .713 | .181 | .530 | 3.937 | .000 | .323 | .365 | .363 |
| descenzh | .031 | .117 | .024 | .260 | .795 | .027 | .026 | .024 |
| xzhicen | .306 | .145 | .283 | 2.105 | .038 | −.102 | .205 | .194 |

a. Dependent Variable: gensat.

**Table 10.5 MLR Output for Simple Slope of attmgrcen with Job Satisfaction When desmgcen Is Low**

| | | | Coefficients[a] | | | | | | |
|---|---|---|---|---|---|---|---|---|---|
| | | Unstandardized Coefficients | | Standardized Coefficients | | | Correlations | | |
| Model | | B | Std. Error | Beta | t | Sig. | Zero-order | Partial | Part |
| 1 | (Constant) | 4.017 | .166 | | 24.251 | .000 | | | |
| | attmgrcen | .108 | .199 | .080 | .544 | .588 | .323 | .054 | .050 |
| | descenzl | .031 | .117 | .024 | .260 | .795 | .027 | .026 | .024 |
| | xzlocen | .306 | .145 | .311 | 2.105 | .038 | .374 | .205 | .194 |

a. Dependent Variable: gensat.

We would also need to create a new product variable because we changed the original, so we would multiply "attmgrcen" by "descenzl" to get "xzlocen." The resulting regression equation where "desmgcen" is a *low value*, shown in Table 10.5, is

$$\text{jobsatis} = (.108)_{\text{attmgrcen}} + 4.017$$

The two separate regression equations above represent equations where "attmgrcen" predicts "gensat"—one where "desmgcen" is high, and one where "desmgcen" is low. We can now plot each of these equations to see how high and low values of the second predictor affect the regression of "gensat" on the first predictor.

In order to plot the equations, we need to substitute high, medium, and low values of "attmgrcen" in both equations. This will result in three data points for each equation. I used Microsoft Excel to create a graph with these calculated values. As you can see from Table 10.6, I reproduced the regression coefficients ($b$ and intercepts) when desmgcen was high and low. The table shows high, medium, and low substitution values of attmgrcen ((+1 SD), 0, and (–1) SD of attmgrcen) to predict gensat for each level. The final column of predicted values can be used to generate a line graph that includes a regression line for desmgcen high and desmgcen low.

As shown in Figure 10.2, there are two regression lines, representing the high and low values of "desmgcen" with three "attmgrcen" data points each. Clearly, when "desmgcen" is low, "attmgrcen" does not have much of an impact on "gensat," which you can see by the relatively flat regression line. However, when "desmgcen" is high, there is a strong positive relationship between "gensat" and "attmgrcen." This illustrates

**Table 10.6   Table for Graphing Simple Slopes of Job Satisfaction Regressed on attmgrcen**

| $b$ | attmgrcen Value Used for Prediction | Intercept | desmgcen High $Y'$ |
|---|---|---|---|
| 0.713 | High (.97) | 4.077 | 4.76861 |
| 0.713 | Medium (0) | 4.077 | 4.077 |
| 0.713 | Low (−.97) | 4.077 | 3.38539 |

| $b$ | attmgrcen Value Used for Prediction | Intercept | desmgcen Low $Y'$ |
|---|---|---|---|
| 0.108 | High (.97) | 4.017 | 4.12176 |
| 0.108 | Medium (0) | 4.017 | 4.017 |
| 0.108 | Low (−.97) | 4.017 | 3.91224 |

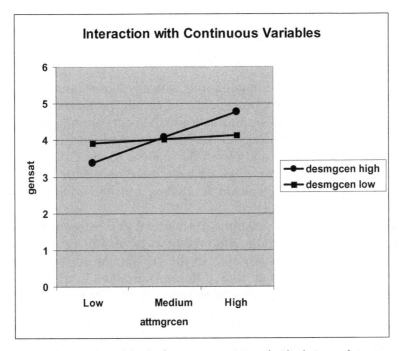

**Figure 10.2**   Interaction of desire for management and attitude toward management in predicting job satisfaction.

the interaction between the two predictors in their relationship to the outcome.

This database included worker job satisfaction, attitudes toward management, and desire for participation in decision-making. The results of this sample of data indicate that attitudes toward management do not relate strongly to job satisfaction when workers have little desire for participation in decision-making. However, when workers definitely want to be involved in making work decisions, attitudes toward management have a positive relationship to job satisfaction.

## INTERACTION WITH CATEGORICAL VARIABLES

Researchers should also test for interaction when categorical variables are included in the regression analysis. There are nuances to interpreting the results of a significant interaction in such cases, but the basics are the same: if there is an interaction, derive equations for different levels of the categorical variable and plot the interaction of these categorical equations, substituting values for the continuous variables in the interaction. You can derive the separate equations from the overall regression equation, and you can easily use SPSS® to derive them.

As with other regression equations with significant interactions, the interacting variables take precedence in the interpretation and further handling of the data. If an interaction is present, you can proceed to plot the interaction as noted above. If there is no interaction, you can delete the product term and re-run the analysis to see if the predictors are significant. Follow the rules we have listed above for hierarchical analysis, checking for significant $R^2$ change with the addition of each predictor.[31]

I have listed below an example of an interaction between a categorical (dummy) variable and a continuous variable. This example draws from the same database we used in the last section (interaction with continuous variables), but with different variables from the overall database of worker participation. In this example I regressed job satisfaction on attitudes toward management ("attmgmt") and membership in worker participation groups ("wpmem" where "1" indicates membership), along with the product term ("wpxattmgr"). I do not use centering with the categorical variable present.

The relevant SPSS® output for the regression is included in Table 10.7. As shown in the table panels, the interaction term is

---

[31] In the example I use below, the relationship between attmngr and satisfaction is in the "same" direction at both levels of desire for management, so it is a little less of a problem to interpret the main effect of attmngr. However, when there is a "crossover" of the two regression lines and the effects switch at different levels of the other predictor, the interpretation is a bit more complex.

**Table 10.7    MLR Output for Interaction with Categorical Variables**

Model Summary

| Model | R | R Square | Adjusted R Square | Std. Error of the Estimate | Change Statistics R Square Change | F Change | df1 | df2 | Sig. F Change |
|---|---|---|---|---|---|---|---|---|---|
| 1 | .406[a] | .165 | .157 | 1.16408 | .165 | 22.275 | 1 | 113 | .000 |
| 2 | .430[b] | .185 | .170 | 1.15495 | .020 | 2.794 | 1 | 112 | .097 |
| 3 | .483[c] | .234 | .213 | 1.12492 | .049 | 7.060 | 1 | 111 | .009 |

a. Predictors: (Constant), attmgmt.
b. Predictors: (Constant), attmgmt, wpmem.
c. Predictors: (Constant), attmgmt, wpmem, wpxattmgr.

Coefficients[a]

| Model | | Unstandardized Coefficients B | Std. Error | Standardized Coefficients Beta | t | Sig. | Correlations Zero-order | Partial | Part |
|---|---|---|---|---|---|---|---|---|---|
| 1 | (Constant) | 1.801 | .499 | | 3.609 | .000 | | | |
| | attmgmt | .556 | .118 | .406 | 4.720 | .000 | .406 | .406 | .406 |
| 2 | (Constant) | 1.651 | .503 | | 3.282 | .001 | | | |
| | attmgmt | .552 | .117 | .402 | 4.715 | .000 | .406 | .407 | .402 |
| | wpmem | .361 | .216 | .143 | 1.672 | .097 | .152 | .156 | .143 |
| 3 | (Constant) | 2.826 | .660 | | 4.282 | .000 | | | |
| | attmgmt | .266 | .157 | .194 | 1.695 | .093 | .406 | .159 | .141 |
| | wpmem | −2.146 | .967 | −.848 | −2.220 | .028 | .152 | −.206 | −.184 |
| | wpxattmgr | .607 | .228 | 1.041 | 2.657 | .009 | .273 | .245 | .221 |

a. Dependent Variable: gensat.

significant (Sig. $F$ Change = .009), adding about 5% to the overall $R^2$. The resulting equation (taken from model 3 of the second panel) is

$$jobsatis = .266_{attmgmt} - 2.146_{wpmem} + .607_{wpxattmgr} + 2.826$$

We could derive separate regression equations for wpmem categories of "1" and "0," but the SPSS® split file option allows us to do this easily. Simply select "Data–Split File–Organize output by groups" and select the dummy variable, in this case, "wpmem." This procedure will produce the following SPSS® split file menu. As Figure 10.3 shows, we can select "Organize output by groups" and specify "wpmem" in the window "Groups Based on."

Splitting the file by the categorical variable creates separate statistical procedures based on the values of the group used in the split.

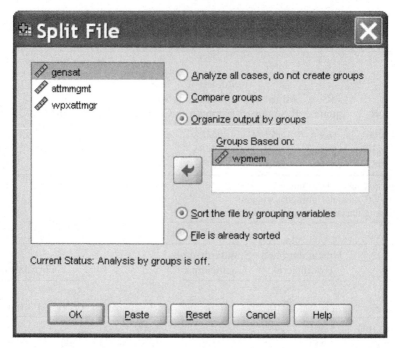

**Figure 10.3** SPSS® panel showing split file specification.

In this example we specified a split file based on the two groups of wpmem ("1" member, and "0" not a member). We can then use the process to run a regression analysis specifying the output variable ("gensat" or satisfaction) regressed on the continuous predictor variable ("attmgmt"). This will create two regression analyses, one for each group of wpmem. Note that the split file procedure will continue to function for any subsequent analysis we request. It is therefore important to remember to "turn off" the split file function when you no longer need this action by reversing the process outlined above; select "Data–Split File" and then select the top option "Analyze all cases, do not create groups."

Once you split the file, you can specify an analysis regressing the outcome variable on only the continuous variable. The SPSS® output in Table 10.8 and Table 10.9 will provide separate results for each of the wpmem groups, as shown below for this example. The resulting equations for both categories of the dummy variable are

$$\text{jobsatis} = .266 \text{ attmgmt} + 2.826 \qquad (\text{where wpmem} = 0)$$

$$\text{jobsatis} = .872 \text{ attmgmt} + .680 \qquad (\text{where wpmem} = 1)$$

**Table 10.8   MLR Split File Output for Categorical Variables: wpmem = 0**

Model Summary[b]

| Model | R | R Square | Adjusted R Square | Std. Error of the Estimate | Change Statistics | | | | |
|---|---|---|---|---|---|---|---|---|---|
| | | | | | R Square Change | F Change | df1 | df2 | Sig. F Change |
| 1 | .204[a] | .042 | .025 | 1.19147 | .042 | 2.560 | 1 | 59 | .115 |

a. Predictors: (Constant), attmgmt.
b. wpmem = .00.

Coefficients[a,b]

| Model | | Unstandardized Coefficients | | Standardized Coefficients | t | Sig. | Correlations | | |
|---|---|---|---|---|---|---|---|---|---|
| | | B | Std. Error | Beta | | | Zero-order | Partial | Part |
| 1 | (Constant) | 2.826 | .699 | | 4.043 | .000 | | | |
| | attmgmt | .266 | .166 | .204 | 1.600 | .115 | .204 | .204 | .204 |

a. wpmem = .00.
b. Dependent Variable: gensat.

**Table 10.9   MLR Split File Output for Categorical Variables: wpmem = 1**

Model Summary[b]

| Model | R | R Square | Adjusted R Square | Std. Error of the Estimate | Change Statistics | | | | |
|---|---|---|---|---|---|---|---|---|---|
| | | | | | R Square Change | F Change | df1 | df2 | Sig. F Change |
| 1 | .618[a] | .381 | .369 | 1.04429 | .381 | 32.050 | 1 | 52 | .000 |

a. Predictors: (Constant), attmgmt.
b. wpmem = 1.00.

Coefficients[a,b]

| Model | | Unstandardized Coefficients | | Standardized Coefficients | t | Sig. | Correlations | | |
|---|---|---|---|---|---|---|---|---|---|
| | | B | Std. Error | Beta | | | Zero-order | Partial | Part |
| 1 | (Constant) | .680 | .656 | | 1.037 | .305 | | | |
| | attmgmt | .872 | .154 | .618 | 5.661 | .000 | .618 | .618 | .618 |

a. wpmem = 1.00.
b. Dependent Variable: gensat.

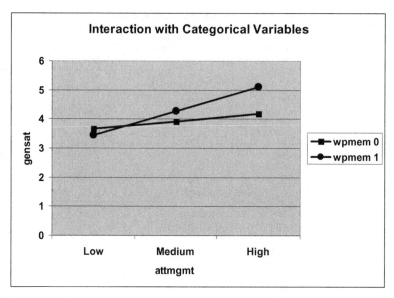

**Figure 10.4**    Interaction graph for categorical variables.

I should mention here that this method is an easy way to derive the regression equations, but there may be slightly different results than using the two-step approach I used earlier. This is the result of the split file process using roughly half of the data, whereas in the earlier approach the simple slopes were tested in the context of all the data. In addition, since this method focuses on all the values of the continuous predictor, you would need to create high, medium, and low values of the continuous variable within the split file process if you want to see the effects of the categorical variable at these levels. The interaction graph that follows (Figure 10.4) identifies these levels of the continuous predictor in order to illustrate the significant interaction.

Once you have the separate regression equations, you can proceed as we did in the last section (interaction with continuous variables) to plot the interaction. Substitute high, medium, and low values of the continuous variable and plot the resulting lines. In Table 10.10, you can see that I reproduced the elements of the regression equations for wpmem = 0 and wpmem = 1 and generated predicted values of job satisfaction in the final column.

Figure 10.4 shows the results of this procedure in which I created a line graph in Microsoft Excel using the predicted job satisfaction values in the final column of the table. As shown in the chart, there is little relationship between attitudes toward management and job

**Table 10.10   Table Used to Generate Graphing Values for Categorical Interaction**

| $b$ | attmgmt Value Used for Prediction | wpmem = 0 | |
|-----|-----------------------------------|-----------|---|
| | | Intercept | $Y'$ |
| 0.266 | High (5.08) | 2.826 | 4.17728 |
| 0.266 | Medium (4.13) | 2.826 | 3.92458 |
| 0.266 | Low (3.18) | 2.826 | 3.67188 |
| $b$ | attmgmt Value Used for Prediction | wpmem = 1 | |
| | | Intercept | $Y'$ |
| 0.872 | High (5.08) | 0.68 | 5.10976 |
| 0.872 | Medium (4.13) | 0.68 | 4.28136 |
| 0.872 | Low (3.18) | 0.68 | 3.45296 |

satisfaction when workers are not members of worker decision-making groups (wpmem = 0). However, decision-making group members (wpmem = 1) indicate stronger job satisfaction, the stronger their attitudes are toward management. Although there were different study variables in this analysis from the analysis in the previous section, the interpretation may be similar for this subset of the overall database: those who are involved in decision-making have higher job satisfaction to the extent that they have stronger attitudes toward management.

## DISCOVERY LEARNING

While not immediately apparent, some predictors are related to outcome variables in such a way that high and low variables of one predictor may differentially affect the action of another predictor variable on the outcome. This interaction effect should be analyzed before other findings, since it is the primary relationship among predictors and outcome variables.

Interaction analyses can include models with categorical as well as continuous variables. Plotting the resulting simple slope equations will help illustrate how the predictors are related to the outcome variable.

Substantively, we saw in two analyses that worker attitudes and participation in decision-making are related to worker satisfaction, but in interactive ways. On one hand, we saw that attitudes toward management do not impact job satisfaction much when workers have little desire for participation in decision-making. However, when workers definitely want to be involved in some work decisions, attitudes toward management have a positive relationship to job satisfaction.

On the other hand, workers differ in work satisfaction, depending on their participation in decision-making groups as well as their attitudes toward management. There is little relationship between attitudes toward management and job satisfaction when workers are not members of worker decision-making groups. However, decision-making group members indicate stronger job satisfaction, the stronger their attitudes are toward management.

Although these findings come from different studies of worker participation, the interpretation may be similar for both studies: those workers who are involved in decision-making or have strong desire for participation have stronger job satisfaction to the extent that they have stronger attitudes toward management.

## Terms and Concepts

**Centering**  A process of subtracting a constant amount from a study variable, typically using the mean of the variable. The result is a set of values that, when summed, will equal zero, or near zero (when using mean centering). The advantage of doing centering is reducing the possibility of multicollinearity among the predictors when there is an interaction, thereby allowing you to have more confidence in the resulting regression estimates. This will reduce the correlations between the product term and the predictor variables and thus reduce multicollinearity. The standard deviations of the centered variables will be unchanged, but the regression estimates will change.

**Disordinal interactions**  Group regression lines that cross within the range of values used to create them.

**Interaction**  Present when the relationship between one predictor and the outcome variable changes at different levels of another predictor variable.

**Ordinal interactions**  Interactions that show separate group regression lines that are not parallel but may not actually meet or "cross" within the range of the data used to create the analysis.

**Product terms**  A variable created by multiplying the values of two variables. Typically these are used to test for an interaction.

**Simple slopes**  A term describing the regression line slope between predictor and outcome at a specific value of a second predictor. (Also referred to as "simple effects" by some authorities.)

## Real World Lab—Interaction

In this exercise you will conduct a MLR analysis using categorical and continuous predictors and a product term. The hypothetical data involve school level findings in which you will predict math achievement from a categorical (Family Income) and a continuous (Classratio) predictor. The database is shown below, and includes the following variables:

- Math achievement ("MathAch")—a continuous outcome.
- Students per classroom ("Classratio")—a continuous predictor.
- Family income ("FamilyIncome3")—a categorical predictor (3 groups) measuring the percentage of students at school with low, medium, and high family income.
- Dummy subvariables for FamilyIncome3: "dsubvar1," "dsubvar2," and "dsubvar3."
- The product variable for the two predictors, which consists of three subvariables:
  S1Clrat—Subvariable 1 × Classratio
  S2Clrat—Subvariable 2 × Classratio
  S3Clrat—Subvariable 3 × Classratio

1. Conduct the MLR analysis.
2. Graph the results showing the interaction (if present).

| MathAch | Classratio | FamilyIncome3 | dsubvar1 | dsubvar2 | dsubvar3 | S1 × Clrat | S2 × Clrat | S3 × Clrat |
|---|---|---|---|---|---|---|---|---|
| 10 | 17 | 1 | 1 | 0 | 0 | 17.00 | 0.00 | 0.00 |
| 11 | 17 | 1 | 1 | 0 | 0 | 17.00 | 0.00 | 0.00 |
| 11 |  | 1 | 1 | 0 | 0 |  | 0.00 | 0.00 |
| 18 | 16 | 1 | 1 | 0 | 0 | 16.00 | 0.00 | 0.00 |
| 18 | 17 | 1 | 1 | 0 | 0 | 17.00 | 0.00 | 0.00 |
| 22 | 19 | 1 | 1 | 0 | 0 | 19.00 | 0.00 | 0.00 |
| 22 | 17 | 1 | 1 | 0 | 0 | 17.00 | 0.00 | 0.00 |
| 24 | 15 | 1 | 1 | 0 | 0 | 15.00 | 0.00 | 0.00 |
| 27 | 17 | 1 | 1 | 0 | 0 | 17.00 | 0.00 | 0.00 |
| 33 | 20 | 1 | 1 | 0 | 0 | 20.00 | 0.00 | 0.00 |
| 46 | 16 | 1 | 1 | 0 | 0 | 16.00 | 0.00 | 0.00 |
| 3 | 13 | 2 | 0 | 1 | 0 | 0.00 | 13.00 | 0.00 |
| 9 | 5 | 2 | 0 | 1 | 0 | 0.00 | 5.00 | 0.00 |
| 18 | 15 | 2 | 0 | 1 | 0 | 0.00 | 15.00 | 0.00 |
| 26 | 14 | 2 | 0 | 1 | 0 | 0.00 | 14.00 | 0.00 |
| 27 | 18 | 2 | 0 | 1 | 0 | 0.00 | 18.00 | 0.00 |
| 34 | 13 | 2 | 0 | 1 | 0 | 0.00 | 13.00 | 0.00 |

| MathAch | Classratio | FamilyIncome3 | dsubvar1 | dsubvar2 | dsubvar3 | S1 × Clrat | S2 × Clrat | S3 × Clrat |
|---|---|---|---|---|---|---|---|---|
| 39 | 15 | 2 | 0 | 1 | 0 | 0.00 | 15.00 | 0.00 |
| 48 | 18 | 2 | 0 | 1 | 0 | 0.00 | 18.00 | 0.00 |
| 48 | 14 | 2 | 0 | 1 | 0 | 0.00 | 14.00 | 0.00 |
| 50 | 19 | 2 | 0 | 1 | 0 | 0.00 | 19.00 | 0.00 |
| 52 | 17 | 2 | 0 | 1 | 0 | 0.00 | 17.00 | 0.00 |
| 52 | 18 | 2 | 0 | 1 | 0 | 0.00 | 18.00 | 0.00 |
| 53 | 16 | 2 | 0 | 1 | 0 | 0.00 | 16.00 | 0.00 |
| 84 | 19 | 2 | 0 | 1 | 0 | 0.00 | 19.00 | 0.00 |
| 4 | 44 | 3 | 0 | 0 | 1 | 0.00 | 0.00 | 44.00 |
| 4 | 20 | 3 | 0 | 0 | 1 | 0.00 | 0.00 | 20.00 |
| 7 | 45 | 3 | 0 | 0 | 1 | 0.00 | 0.00 | 45.00 |
| 9 | 22 | 3 | 0 | 0 | 1 | 0.00 | 0.00 | 22.00 |
| 18 | 11 | 3 | 0 | 0 | 1 | 0.00 | 0.00 | 11.00 |
| 19 | 19 | 3 | 0 | 0 | 1 | 0.00 | 0.00 | 19.00 |
| 24 | 11 | 3 | 0 | 0 | 1 | 0.00 | 0.00 | 11.00 |
| 25 | 16 | 3 | 0 | 0 | 1 | 0.00 | 0.00 | 16.00 |
| 36 | | 3 | 0 | 0 | 1 | 0.00 | 0.00 | |
| 37 | 17 | 3 | 0 | 0 | 1 | 0.00 | 0.00 | 17.00 |
| 40 | 19 | 3 | 0 | 0 | 1 | 0.00 | 0.00 | 19.00 |
| 54 | 18 | 3 | 0 | 0 | 1 | 0.00 | 0.00 | 18.00 |
| 56 | 19 | 3 | 0 | 0 | 1 | 0.00 | 0.00 | 19.00 |
| 56 | 18 | 3 | 0 | 0 | 1 | 0.00 | 0.00 | 18.00 |
| 57 | 17 | 3 | 0 | 0 | 1 | 0.00 | 0.00 | 17.00 |
| 61 | 16 | 3 | 0 | 0 | 1 | 0.00 | 0.00 | 16.00 |
| 63 | 16 | 3 | 0 | 0 | 1 | 0.00 | 0.00 | 16.00 |
| 68 | 17 | 3 | 0 | 0 | 1 | 0.00 | 0.00 | 17.00 |
| 70 | 20 | 3 | 0 | 0 | 1 | 0.00 | 0.00 | 20.00 |
| 78 | 16 | 3 | 0 | 0 | 1 | 0.00 | 0.00 | 16.00 |

# 11

# DISCOVERY LEARNING THROUGH CORRELATION AND REGRESSION

## OVERALL DISCOVERY NOTES

In examining statistical methods that evaluation researchers use, we have explored how to bring new insights to evaluation questions. Analyzing data in appropriate ways provides evaluators confidence that their work of discovering hidden facets of difficult problems can bring the evaluator closer to creating meaningful solutions.

However, it is important to remember in the final analysis that data, and interpretations made from the analyses of data, can be problematic. Data are almost always flawed, due to their nature and level, the way they were collected, and how they are managed. These flaws limit the applicability of the findings based on the data. Further, program sponsors often make use of the analyses in ways the evaluator did not intend or suggest. In both cases, the meaningfulness and usefulness of the evaluator's discoveries are subject to the larger contexts within which evaluation takes place.

Correlation and regression methods like those we explored can be used to illuminate and discover patterns of meaning not immediately apparent from descriptive analyses. They do not guarantee that the results of the analyses will be used properly, however. But, once illuminated, the patterns may challenge previously held assumptions about

*The Program Evaluation Prism: Using Statistical Methods to Discover Patterns*,
by Martin Lee Abbott
Copyright © 2010 John Wiley & Sons, Inc.

the nature of the problems prompting the collection of the data. Practitioners who attempt to create improved achievement or worker outcomes should recognize that their assumptions about available data are more complex because of the settings in which they are found.

## FINDINGS FROM THE DATA

### Student Academic Achievement

You may recall that earlier in the book I described the achievement gap and concerns over how to manage such a difficult and intractable problem. While the gap is real, the solution may be tied more to understanding the full nature of the problem rather than making premature assumptions. If policy makers and school leaders make a judgment about what needs to be done in the absence of continued data analyses, especially under great popular and legislative pressure, they might miss the real determinant(s). This is a deeply embedded problem, and there are no magic bullets. Many concerned groups and granting bodies have taken great pains and spent billions of dollars to redress what has become part of the landscape of American education. Many innovative programs are underway to make inroads into this and related problems among K–12 education. It is fair to say that whatever solution that emerges will likely involve thorough re-institutionalizing of our schools and learning paradigms. It will not be quick or easy.

These matters go far beyond the scope of this book, but I mention them because I have used a demonstration database with some of the variables that embody the substance of this controversy. What I have advocated in this book is that we must be fully aware of the extent of a problem before we have a hope of solving it. Therefore, what other aspects of the achievement gap might not be apparent, but nevertheless real counteracting forces to positive change?

The power of correlation methods can help us understand this complex reality beyond what some data appear to indicate, or beyond what some powerful social groups assume to be true. Multiple regression is based on correlation and helps to minimize "interference" from outside influences so that you can see the unique influence of one variable upon an outcome. We can use this method to see what variables influence achievement and how each contributes to an understanding of the whole.

In the data analyses we have presented, family income overshadows several other influences on student academic achievement. Of course, the nature of our findings could be explained partially by the nature of our data. We only had access to aggregated school findings, which meant that we had measures of economic standing, ethnicity, and other variables for the entire student populations of the schools and not for individual students. Further we used the data, albeit "real" data, as an example of how information can be gleaned from general analyses. Therefore we could not make definitive conclusions about the relationships among ethnicity, achievement, and economic standing for individual students, for example.

The analyses we have considered do not refute the existence of an achievement gap, but rather indicate that the nature and reasons for the gap are not fully understood. Careful use of available statistical tools can help illuminate facets of the relationships not previously recognized. We will still be challenged by the lack of information about other potentially important influences that, if they were taken into account, could help further illuminate a crucially important part of American education.

As I mentioned at the outset of this book, evaluation research often takes place in a contested political climate. The evaluator faces the challenge of providing the most objective results with the data available. The sponsor of the research may use the data in various ways. In some research projects, evaluators are in a position to make policy recommendations based on the findings of their work. At this point their role can subtly change from objective researcher to one giving direction to the program being evaluated. As an example, the findings about ethnicity and family income might have implications for what resources are made available for students in schools. Or, the findings might suggest that other research would help to refine programs that might lead to "better" achievement outcomes, like considering class size in the analyses. The evaluator must always balance the discovery of the findings with the way in which they are used by others.

## Workplace Participation

Our examination of the worker participation database also was instructive as we applied the multiple regression method to help acquire insights not immediately apparent from the data. In this case the data were at the individual level, since they were gathered through survey methods and interviews. The primary limitation with this database was the lack of generalizability. Because the data were taken from one site that was not selected by random means, no conclusions could be drawn that would apply to similar workers in similar plants in similar workplace participation programs. At best, we could treat the data as illustrating how multiple regression works, and to make descriptive statements about this setting that might be helpful to further research of the same kind. It is also illustrative of the kind of data with which evaluators often must be content as they seek to address specific, localized problems.

Within the context of these limitations, we found some interesting insights. One of these insights was that simply having a mechanism to provide workers an outlet for decision-making was not necessarily alone effective as an influence on worker satisfaction. As we found in a couple of different studies, there appeared to be an interaction between workplace participation and worker attitudes as they together accounted for worker satisfaction. As we stated earlier, those who are involved in decision-making or have strong desire for participation have higher job satisfaction to the extent that they have stronger attitudes toward management.

One implication of these findings is that program supervisors need to make sure they understand the attitudes of their workers before they create workplace participation opportunities. Just having the programs is not enough. The programs must operate fairly and be meaningful avenues to real workplace change. They might also serve as a way to identify workers whose attitudes toward the management role could be useful in the overall functioning of the work.

## Impact of Technology on Student Learning

I used TAGLIT data several times in the book to illustrate the use of regression, correlation, and related procedures. These data have proved very helpful to national efforts to understand the nature and depth of teacher skills and use of technology vis-à-vis their students. The data we used did not include a student achievement variable for several reasons. First, the data were nationwide, with each state providing their own achievement measures. Because we had the benefit of individual-level teacher and student data, we would have had to aggregate the results to school levels, and then find a measure that equated the different state tests. Second, the focus of the studies were on the use of technology and the effects on the classroom. This was an important focus prior to understanding the extension of these results to achievement, which is a tenuous analysis. It was more important early in the studies to try to understand such things as: What discrepancies in technology skill and use there were between teacher and student? To what extent was technology used in the classroom? What access and support were available to schools and teachers for doing so? What impact did technology have on teaching and learning activities in the classroom?

The study and findings using all these data go beyond the nature of this book, since we are interested in the methods used to understand the data. However, from the glimpses of our analyses of the early TAGLIT data in this book, it was readily apparent in our studies that any impact on classroom teaching and learning was strongly related to the technology skill of teachers and students. To be sure, teachers needed support and access to technology, but whether it was used effectively depended on their understanding of the technology available. Issues of family income and ethnicity emerged as being small additional influences, but knowledge of technology appeared to be the most important determinant of teaching and learning as it was measured by the TAGLIT instruments.

## ADVANCED STATISTICAL TECHNIQUES

Thus far we have explored multiple regression using categorical and continuous variables. The results of these analyses, which were for the most part based on actual research studies, provided some insight into how to interpret data output and into the nature of the research topics. Evaluators will see, however, that there are further statistical applications and procedures that can go much further to assist with interpretations of difficult evaluation questions. In what follows I will briefly summarize some of these methods.

### Hierarchical Linear Modeling

This method is very helpful to researchers who deal with data that are found in different "levels." One example is educational data that are gathered on the student level but are also "nested" within other analytical levels, like classrooms, or different size schools. We made brief mention of this issue when we introduced our education database that was aggregated to the school level. If we had access to individual student results, it would have been important to conduct analyses whereby we could regress individual student academic results on student family income level, and ethnic group, for example, but include classroom size, teacher's years of experience, as well as other variables that measure different analytical levels (school size, district size, etc.).

There are many advantages to doing this kind of analysis, variously called *hierarchical linear modeling* (HLM) or *multilevel regression,* since it takes into account multiple levels of data. Using such methods, one can see what relationship exists among variables on the student level, as well as how these relationships are impacted by the "larger" levels of data. In particular, we could see what interactions exist between the first and second (and beyond) levels of data. This information is very helpful to the overall interpretation because it takes into account all the information available in the same analysis. The most comprehensive treatment of hierarchical linear models is found in the work of Raudenbush and Bryk (2002). Their stand-alone software application (Raudenbush et al., 2004) is available for these kinds of analyses, and should be considered for data that are nested.

To provide one example, at the Washington School Research Center I worked with two colleagues to replicate a study by Bickel and Howley (2000) in which the researchers used a multilevel approach in their analysis of the joint influence of district and school size on academic performance in Georgia. As part of that study, my colleagues and I (Abbott et al., 2002) conducted a multilevel analysis for understanding the influence of district size, school size, and socioeconomic status on student achievement in Washington. Because these data are from different analytical levels, we needed a method that would recognize the nested nature of the data and identify the effects of the "larger" or second level of data (e.g., district size) on achievement and other school-level characteristics (e.g., family income). Using HLM, we found, among other things, that large district size was detrimental to achievement (4th and 7th grades) in that it strengthens the negative relationship between school poverty and student achievement. Without HLM, we could not have understood the additional influence on the achievement–family income relationship represented by district level size.

There are several good resources for understanding multilevel analyses. Earlier, I cited Raudenbush and Bryk (2002), but a very good more recent work by Bickel (2007) provides a thorough and approachable treatment of this interesting procedure.

## Structural Equation Modeling and Path Analysis

Other important procedures that are an extension of the topics in this book are *structural equation modeling* and *path analysis*. I will not elaborate on these procedures here, but will only mention that they allow the evaluator to come a bit closer to examining the causal relationships among study variables by understanding the direct and indirect relationships.

In Chapter 8 on multiple correlation, I defined path analysis and structural equation modeling as attempts to understand causal relationships using visual models. Both procedures assume a theoretically driven framework for understanding the relations among study variables, with the latter (SEM) being a more comprehensive way to identify the relationships among both observed and unobserved variables. Path analysis is a more limited procedure that focuses on the direct and indirect relationships among a set of observable variables.

An example of path analysis comes from work my colleagues and I conducted on the relations among several predictor variables and school achievement outcomes. Figure 11.1 shows the type of paths we used in our path analysis.

As the figure shows, you can measure how ethnicity directly affects math achievement (a direct arrow or "path" is not included in the model to show this), and you can also see that ethnicity is related "indirectly" to math achievement through its influence on the two SES measures, and then through a series of individual student behaviors.

Path analysis allows the evaluator to measure each of the direct and indirect impacts of all the relationships in the model through the use of the standardized betas produced in a series of MLR analyses. We touched on these analyses in this book in our analyses of how ethnicity and family income relate to school-level achievement. As you may recall, we noted that ethnicity is related to math achievement, but primarily through its relationship to the school's proportion of lower income students.

## OTHER REGRESSION PROCEDURES

There are many other important regression methods, one of which is used heavily in sociological research. *Logistic regression* is a regression method used with categorical outcomes. Most commonly it uses

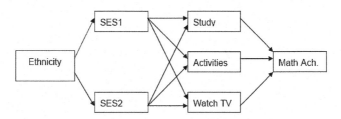

**Figure 11.1**    SEM chart example.

predictors that estimate the probability of a binomial outcome variable, or one that has two possibilities such as "yes–no," "graduated–not graduated," and so on. *Multinomial logistic regression* uses the same procedure for estimating the probability of a categorical variable with more than two possible outcomes (e.g., "symptom retention, reversal, or substitution" in therapeutic contexts). A good resource to begin your study of logistic regression is Hosmer and Lemeshow (2000).

Beyond these are MLR-based procedures of *discriminant analysis* and *factor analysis*. If you recall, I introduced factor analysis in Chapter 4 on correlation when I reported on a TAGLIT study that focused on identifying a few factors "distilled" from a long questionnaire with multiple items among middle/high school teachers.

If your interest in statistical methods has increased by considering our topics thus far, you may wish to explore these advanced methods. Evaluators who explore these and other methods will be well prepared to discover patterns in their data and prepare studies that are more meaningful. Brown (2006) and McLachlan (1992) are some resources to consider for the additional works.

# 12

# PRACTICAL APPLICATION ANALYSES

## REAL WORLD LAB—CORRELATION

Conduct a correlation analysis among the variables in the data set using SPSS®. Examine the correlation matrix and respond to these questions:

1. Are any of the correlations significant?

Answer: The following correlations are statistically significant (see Table 12.1):
   Reading–Math ($p < .000$)
   Reading–PctFR ($p < .032$)
   Reading–Star_mean ($p < .027$)
   Math–PctFR ($p < .009$)

*The Program Evaluation Prism: Using Statistical Methods to Discover Patterns*,
by Martin Lee Abbott
Copyright © 2010 John Wiley & Sons, Inc.

**Table 12.1  Correlation Matrix with School Variables**

<table>
<tr><td colspan="6" align="center">Correlations</td></tr>
<tr>
<td></td>
<td></td>
<td align="center">Reading<br>PercentMet<br>Standard</td>
<td align="center">Math<br>PercentMet<br>Standard</td>
<td align="center">Percent<br>FreeorReduced<br>PriceMeals</td>
<td align="center">STAR_<br>mean</td>
</tr>
<tr>
<td>Reading<br>PercentMet<br>Standard</td>
<td>Pearson<br>Correlation<br>Sig. (2-tailed)<br>N</td>
<td>1<br><br><br>32</td>
<td>.734**<br><br>.000<br>32</td>
<td>−.392*<br><br>.032<br>30</td>
<td>.392*<br><br>.027<br>32</td>
</tr>
<tr>
<td>Math<br>PercentMet<br>Standard</td>
<td>Pearson<br>Correlation<br>Sig. (2-tailed)<br>N</td>
<td>.734**<br><br>.000<br>32</td>
<td>1<br><br><br>32</td>
<td>−.466**<br><br>.009<br>30</td>
<td>.139<br><br>.447<br>32</td>
</tr>
<tr>
<td>PercentFreeor<br>Reduced<br>PriceMeals</td>
<td>Pearson<br>Correlation<br>Sig. (2-tailed)<br>N</td>
<td>−.392*<br><br>.032<br>30</td>
<td>−.466**<br><br>.009<br>30</td>
<td>1<br><br><br>30</td>
<td>.038<br><br>.844<br>30</td>
</tr>
<tr>
<td>STAR_mean</td>
<td>Pearson<br>Correlation<br>Sig. (2-tailed)<br>N</td>
<td>.392*<br><br>.027<br>32</td>
<td>.139<br><br>.447<br>32</td>
<td>.038<br><br>.844<br>30</td>
<td>1<br><br><br>32</td>
</tr>
</table>

\*\*Correlation is significant at the 0.01 level (2-tailed).
\*Correlation is significant at the 0.05 level (2-tailed).

2. What is the coefficient of determination, or effect size, of the relationship between reading achievement and the STAR_mean?

Answer: $R^2 = .15$ ($.392^2$)

3. Create a scatter diagram of reading achievement as the outcome variable and STAR_mean as the predictor variable. Describe the resulting pattern of the data shown in Figure 12.1.

Answer: The data are widely scattered around the line in an upward direction, indicating a positive correlation. As the values of the STAR_ mean variable increase, the reading achievement values generally increase. Because the data are scattered so widely around the line, we expect that the correlation will be weak, despite the fact that the correlation between the two variables are statistically significant.

4. Express the correlation results in the language of the research problem.

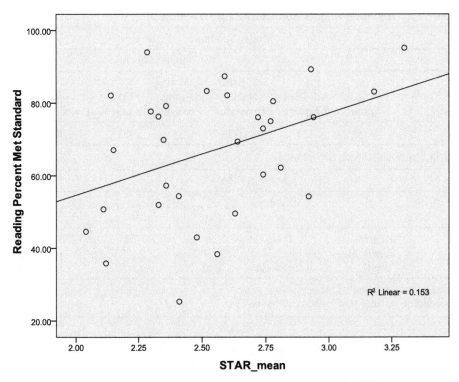

**Figure 12.1**  Bivariate scatter diagram of achievement and STAR mean.

Answer: There is a weak but statistically significant relationship between the observation results and school level reading achievement. As Powerful Teaching and Learning™ results increase among elementary and middle schools in this sample, the schools indicate greater reading achievement results. According to the coefficient of determination, about 15% of the variance in the school reading achievement results are accounted for by the STAR_mean observation values.

## REAL WORLD LAB—BIVARIATE REGRESSION

Use SPSS® to conduct a bivariate regression with the data presented in Chapter 4 on correlation. Regress the outcome variable (Reading Achievement) on the predictor variable (STAR_mean):

The answers to the questions are based on the following SPSS® output (see Table 12.2):

**Table 12.2 Bivariate Regression Output for STAR Data**

Model Summary

| | | | | | Change Statistics | | | | |
|---|---|---|---|---|---|---|---|---|---|
| Model | R | R Square | Adjusted R Square | Std. Error of the Estimate | R Square Change | F Change | df1 | df2 | Sig. F Change |
| 1 | .392[a] | .153 | .125 | 16.88713 | .153 | 5.433 | 1 | 30 | .027 |

a. Predictors: (Constant), STAR_mean.

ANOVA[b]

| Model | | Sum of Squares | df | Mean Square | F | Sig. |
|---|---|---|---|---|---|---|
| 1 | Regression | 1549.290 | 1 | 1549.290 | 5.433 | .027[a] |
| | Residual | 8555.255 | 30 | 285.175 | | |
| | Total | 10104.545 | 31 | | | |

a. Predictors: (Constant), STAR_mean.
b. Dependent Variable: ReadingPercentMetStandard.

Coefficients[a]

| Model | | Unstandardized Coefficients | | Standardized Coefficients | | |
|---|---|---|---|---|---|---|
| | | B | Std. Error | Beta | t | Sig. |
| 1 | (Constant) | 9.442 | 24.886 | | .379 | .707 |
| | STAR_mean | 22.598 | 9.695 | .392 | 2.331 | .027 |

a. Dependent Variable: ReadingPercentMetStandard.

1. What is the overall $R^2$?

Answer: Overall $R^2$ = .153 (the adjusted $R^2$ = .125 and may be preferable in this study due to the small sample size). Note that the overall $R$ value of .392 is the correlation between the predictor and outcome when there is only one predictor in the analysis. (Compare to the results for the correlation problem above.)

2. Is the "omnibus" finding significant?

Answer: the omnibus test (as shown in the "Anova" results section) is statistically significant, $F$ = 5.433, $p < .027$. Since there is only one predictor, the $F$ results will also be reported in the "Model Summary" section under "F Change" and "Sig. F Change." When there is more than one predictor in the model, the model summary reports the change to the overall $F$ value when additional variables are added to the analysis.

3. Is the test of the predictor variable significant?

Answer: The significance test for the predictor is statistically significant as indicated by the $t$-test results in the "Coefficients" panel. The $t$-test resulted in a value of 2.331, which was significant at $p < .027$ (reported in the "Sig." column). Note that the $t$ value of 2.331 when squared equals the $F$ value ($2.331^2 = 5.433$). As additional predictors are added to the analysis, the regression output file provides information for the contribution of each predictor to the $F$ value.

4. Use the sum of squares in the ANOVA results to calculate $R^2$

Answer: Note the sum of square values reported in the Anova table:
Regression (between) sum of squares = 1549.29
Residual (within) sum of squares = 8555.255
Total sum of squares = 10104.545
Recall that Recall that $R^2 = \dfrac{SS(reg)}{SS(Y)}$
Therefore $R^2 = 1549.29/10104.545 = .153$

5. What is the regression formula?

Answer: The (unstandardized) formula: $Y' = 22.598X + 9.442$
The (standardized) formula: $Z_Y = .392Z_X$

6. Provide a scatter diagram of the results.

Answer: The scatter diagram below (Figure 12.2) shows the results of the regression analysis. Note the differences between this graph and the one above (under the correlation problem results). It appears different because the scale of the independent variable ($x$ axis) is different. The correlation graph above shows the $x$ axis values roughly between 2 and 3.25, which are the actual lowest and highest values. However, this is a slight distortion of the data, since the possible values are between 0 and 4. When I adjusted the scale of both axes, the scatter diagram shown resulted. (You can make these adjustments by double-clicking on the scatter diagram that is reported in the SPSS® output file, then double-click on the area of the diagram you want to change, and use the appropriate drop-down menus to specify

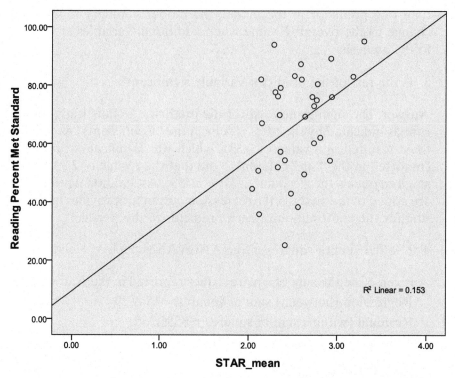

**Figure 12.2** Bivariate scatter diagram with rescaled STAR values.

values for the axes other than the default values.) The scatter diagram shows the same cases, but the relationship appears quite differently. Nevertheless, the same regression values result from the analysis. Note that on the scatter diagram, the $R^2$ value is reported in the lower right corner of the diagram. When these adjustments are made, you can see that the "$y$-intercept" value of 9.442 (from the unstandardized equation above) is positioned correctly in the scatter diagram.

7. Express the regression results in the language of the research problem.

Answer: The STAR_mean significantly predicts the reading achievement level ($F = 5.433$, $p < .027$) in this sample group of schools ($N = 32$). The scatter diagram shows that the two variables are related positively and linearly, although the size of the correlation and the small sample size result in a wide scatter pattern. Therefore, as the STAR_mean values increase among the schools, the school-level reading achievement scores increase. The $R^2 = .153$ (adjusted

$R^2 = .125$) of indicates a small-to-medium effect size of the relationship between school-level reading achievement and Powerful Teaching and Learning™ mean observation values. Approximately 15% of the variance in school level reading achievement is accounted for by STAR_mean values in this sample of schools.

8. Is the relationship linear?

Answer: The following SPSS® output (Table 12.3) shows the results of the curve fit analysis of this regression analysis:

**Table 12.3   Curve Fit Output for Regression Analysis**

| | Model Summary and Parameter Estimates | | | | | | | |
|---|---|---|---|---|---|---|---|---|
| | Dependent Variable: ReadingPercentMetStandard | | | | | | | |
| | Model Summary | | | | | Parameter Estimates | | |
| Equation | R Square | F | df1 | df2 | Sig. | Constant | b1 | b2 |
| Linear | .153 | 5.433 | 1 | 30 | .027 | 9.442 | 22.598 | |
| Quadratic | .163 | 2.822 | 2 | 29 | .076 | 111.261 | −56.449 | 15.119 |

The independent variable is STAR_mean.

According to the results, the linear model is significant ($p < .027$), while the quadratic model is not ($p < .076$). There is a slightly higher $R^2$ with the quadratic model, but this may be an effect of the small sample size. The comparison of the two models is shown in the scatter diagram of Figure 12.3.

## REAL WORLD LAB—EXTREME SCORES

We will continue to explore the relationship between reading achievement and STAR_mean values. In this exercise you will use the STAR Protocol database that you used for the last two practical application exercises. You will attempt to locate outlier scores in the database and discuss the results of your analysis.

Use the SPSS® command to "Analyze–Descriptive Statistics–Descriptives" to save standardized (Z-score) values to identify univariate outliers:

1. Are any of the Z-score values beyond +/– 3.00?

**Reading Percent Met Standard**

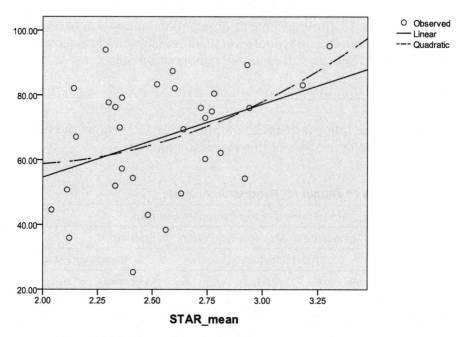

**Figure 12.3**   Curve fit analysis of bivariate regression variables.

Answer: There are no *Z* scores exceeding +/– 3.00. The largest *Z* score for STAR_mean is 2.40262, the top score in Table 12.4 below. The table represents the first 4 cases and shows the standardized values for STAR_mean and readstd (ZSTAR_mean, Zreadstd):

**Table 12.4   Standardized Case Analysis of STAR Data**

| readstd | percfrl | STAR_mean | ZSTAR_mean | Zreadstd |
|---------|---------|-----------|------------|----------|
| 95.20 | 96.78 | 3.30 | 2.40262 | 1.56041 |
| 83.10 | 40.72 | 3.18 | 2.01904 | 0.89020 |
| 76.10 | 66.34 | 2.94 | 1.25188 | 0.50248 |
| 89.30 | 69.40 | 2.93 | 1.21992 | 1.23361 |

2. Visually inspect a scatter diagram to identify any extreme scores and determine whether or not they are "close to the pack."

Answer: The scatter diagram of Figure 12.4 shows the data arrayed around the regression line. The pattern of the cases is a "vertical cluster" but generally tending upward to the right. The $R^2 = .15$ is

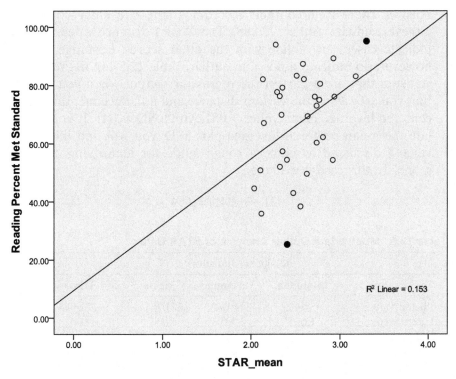

**Figure 12.4** Bivariate analysis showing extreme scores.

indicated in the lower right portion of the graph. There are a couple of cases that may affect the direction of the scatter, indicated by a solid black dot on the graph. We can look at the SPSS® data to further inspect the data.

3. What is your plan with respect to any identified outliers?

Answer: Without a full outlier analysis, we would leave the cases in the database. We might go back to the original data to determine whether the cases might be extreme due to data entry error, or whether the data represented a valid response.

Conduct a regression analysis predicting reading achievement from STAR_mean scores and save the following as variables added to your database to help identify *multivariate outliers*: standardized and unstandardized residuals, DfBeta(s) and standardized DfBeta(s), Cook's distance values, and leverage values.

4. What multivariate outliers can you identify?

Answer: There are no outliers according to the $Z$ residual scores. The largest residual score is (–2.288). The Cook's distance values do not indicate cases out of line with the other scores. Leverage values, however, do indicate a possible outlier. Table 12.5 reports residuals statistics that are a part of the regression output when you specify that you are saving the various distance and influence measures. The centered leverage mean value = .031, with a SD = .041. If you calculate the mean centered leverage plus 3SD, you will find the cutoff value I discussed as being a rough guide for identifying extreme scores. In this case it would be

$$.031 + 3(.041) = .154$$

**Table 12.5   Multivariate Outlier Analysis of STAR Data**

| Residuals Statistics[a] | | | | | |
|---|---|---|---|---|---|
| | Minimum | Maximum | Mean | Std. Deviation | N |
| Predicted Value | 55.5405 | 84.0134 | 67.0281 | 7.06945 | 32 |
| Std. Predicted Value | –1.625 | 2.403 | .000 | 1.000 | 32 |
| Standard Error of Predicted Value | 2.987 | 7.875 | 4.065 | 1.156 | 32 |
| Adjusted Predicted Value | 55.5076 | 80.9521 | 66.9429 | 6.91617 | 32 |
| Residual | –38.64144 | 32.90699 | .00000 | 16.61252 | 32 |
| Std. Residual | –2.288 | 1.949 | .000 | .984 | 32 |
| Stud. Residual | –2.332 | 2.003 | .002 | 1.011 | 32 |
| Deleted Residual | –40.14277 | 34.78490 | .08525 | 17.57271 | 32 |
| Stud. Deleted Residual | –2.534 | 2.116 | –.003 | 1.041 | 32 |
| Mahal. Distance | .001 | 5.773 | .969 | 1.267 | 32 |
| Cook's Distance | .000 | .115 | .029 | .035 | 32 |
| Centered Leverage Value | .000 | .186 | .031 | .041 | 32 |

a. Dependent Variable: ReadingPercentMetStandard.

One centered leverage value in Table 12.6 exceeds this cutoff value of .154. The largest (LEV_1) value is .18621, when the readstd = 95.20 and the STAR_mean = 3.30. This is the case that we identified in the former scatter diagram as being in the upper right part of the scatter. Recall that a case might stand apart from the cluster but not unduly affect the results if it is in the line of the scatter trajectory. In some situations, however, the case may unduly exert an influence on the regression line.

**Table 12.6   Multivariate Case Analysis with STAR Data**

| readstd | STAR_mean | LEV_1 | DFB0_1 | DFB1_1 |
|---------|-----------|-------|--------|--------|
| 95.20 | 3.30 | 0.18621 | −8.57843 | 3.54156 |
| 83.10 | 3.18 | 0.13150 | −1.07244 | 0.44718 |
| 44.60 | 2.04 | 0.08518 | −5.67400 | 2.07469 |
| 50.80 | 2.11 | 0.06334 | −2.78924 | 1.00890 |
| 35.90 | 2.12 | 0.06048 | −9.23433 | 3.33406 |

5. What, if any, impact is there for deleting problematic outliers?

Answer: Because of the nature of this one case, I might further explore the impact on the regression line if I were to delete the case. I can use the "Dfbeta results (shown in Table 12.6) to assist in this analysis.

Recall that the regression equation was

$$Y' = 22.598X + 9.442$$

If we drop the case with an extreme Centered Leverage value, what would be the impact on the regression equation? Subtracting the values under "DFB0_1" and "DFB1_1" will help to assess this impact for the intercept and slope values, respectively. Using the values in Table 12.6 above, the impact on the regression coefficients would be:

$$\text{Slope} = 22.598 - 3.54156 = 19.056$$

$$\text{Intercept} = 9.442 - (-8.57843) = 18.02$$

The resulting regression equation is

$$Y' = 19.056X + 18.02$$

If you delete the one case under consideration and re-run the regression analysis, you would find that this revised regression equation would be the result. That is the value of the Dfbeta values. You can create the new regression equation without deleting and re-running the analysis.

By re-running this analysis (after we deleted the case), we would observe the changes to the $R^2$ value as well. The scatter diagram of Figure 12.5 shows the revised pattern. Notice that the $R^2$ value is now .096 (down from .153 as shown in the previous regression analysis). The effect of dropping one extreme case on the $R^2$ is therefore

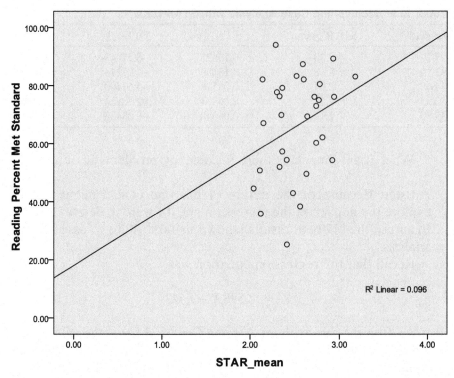

**Figure 12.5**   Scatter diagram showing effects of dropping extreme case.

$$.057 = .153 - .096$$

6. What is your overall plan to deal with multivariate outliers?

Answer: This example demonstrates the dilemma for the evaluation researcher. Often the evaluator has limited data from which to make a decision that may affect policy. In such cases you can see that any changes to the data can strongly affect the outcome. As a further example, you will recognize that even among the first few cases (as shown in the table above) the Dfbeta values indicate a large impact on the regression coefficients. That is typical with small data sets. It is therefore very important when going in to a research project to make sure that the data available are valid and reliable. With small data sets, the evaluator would want to make conclusions carefully and point out the potential difficulties with limited data. In the case we have just examined, the evaluator would want to consider carefully before dropping cases, especially if there is no evidence that the case is not a valid response (rather than reported in error).

# REAL WORLD LAB—PARTIAL AND SEMI-PARTIAL CORRELATION

We discussed partial correlations in Chapter 7 using the school-level data correlating reading and math achievement with family income and ethnicity. The database is a sample of the data that contains these variables. Use these data to respond to the questions below.

Create a correlation matrix and examine the relationships among the study variables:

1. What are the results of the bivariate correlations? Do they match the analyses discussed in Chapter 7 on the partial correlation?

Answer: The bivariate correlations generally match those presented in Chapter 7. The size of the correlations are a bit different due to sample sizes. Generally, math and reading achievement are inversely correlated to family income (percent FreeorReducedPricedMeals) and school-level ethnicity (PercentWhite). All the correlations are statistically significant, as shown in Table 12.7 below.

**Table 12.7  Correlation Matrix for Partial Correlation Analysis**

Correlations

| | | Freeor Reduced PricedMeals | Math Percent MetStandard | Reading Percent MetStandard | Percent White |
|---|---|---|---|---|---|
| Freeor Reduced PricedMeals | Pearson Correlation | 1 | $-.440^*$ | $-.495^{**}$ | $-.717^{**}$ |
| | Sig. (2-tailed) | | .015 | .005 | .000 |
| | N | 30 | 30 | 30 | 30 |
| Math Percent MetStandard | Pearson Correlation | $-.440^*$ | 1 | $.864^{**}$ | $.451^*$ |
| | Sig. (2-tailed) | .015 | | .000 | .012 |
| | N | 30 | 30 | 30 | 30 |
| Reading Percent MetStandard | Pearson Correlation | $-.495^{**}$ | $.864^{**}$ | 1 | $.503^{**}$ |
| | Sig. (2-tailed) | .005 | .000 | | .005 |
| | N | 30 | 30 | 30 | 30 |
| PercentWhite | Pearson Correlation | $-.717^{**}$ | $.451^*$ | $.503^{**}$ | 1 |
| | Sig. (2-tailed) | .000 | .012 | .005 | |
| | N | 30 | 30 | 30 | 30 |

*Correlation is significant at the 0.05 level (2-tailed).
**Correlation is significant at the 0.01 level (2-tailed).

2. Conduct partial correlation analyses on the variables as we did in the chapter discussion. What is the impact on the relationship between reading achievement and school-level ethnicity when family income is removed from both study variables?

Answer: When family income is removed from both the reading achievement and ethnicity variables, the original relationship is no longer significant. Table 12.8 below shows this result. The original bivariate relationship was $r = .503$, $p < .005$. When family income is removed from both variables, the correlation drops to .244, and is no longer statistically significant $p < .20$). The $R^2$ value drops from .25 ($.503^2$) to .06 ($.244^2$).

**Table 12.8   Partial Correlation Output for Ethnicity, Achievement, and Low Income**

| Correlations | | | Percent White | ReadingPercent MetStandard |
|---|---|---|---|---|
| Control Variables | | | | |
| FreeorReduced PricedMeals | PercentWhite | Correlation | 1.000 | .244 |
| | | Significance (2-tailed) | — | .202 |
| | | df | 0 | 27 |
| | ReadingPercent MetStandard | Correlation | .244 | 1.000 |
| | | Significance (2-tailed) | .202 | — |
| | | df | 27 | 0 |

3. Interpret the results of your analyses. What do the results of the partial correlation analyses indicate?

Answer: The partial correlation results indicate that the apparent relationship between reading achievement and ethnicity may be at least partially a function of the correlation of both variables with family income. School-level ethnicity may appear to be a major determinant of reading achievement, but these results may be misleading. The analyses show that both achievement and ethnicity are related to family income, which may be a more meaningful explanation of school-based reading achievement in this sample of schools.

## REAL WORLD LAB—MULTIPLE REGRESSION

In this exercise you will use a sample of the 2003 TAGLIT database that I used for the example in this chapter (the database is included in the Practical Application section of Chapter 8 on the MLR). This time you will regress "mhimpact" (the outcome variable) on two predictors: "mhskills" (the index of teacher computer skills) and "studtch" (a school-level measure of the class size, or student–teacher ratio). Enter the predictors in separate steps so that you can see the change in $R^2$ as each predictor is added to the model. See Table 12.9

1. Is the overall $F$ value for the model significant? (omnibus test)

Answer: The omnibus test for the first model is significant ($F = 5.178$, $p < .028$). However, the omnibus test for the second model, which includes both predictors, is not significant ($F = 2.601, p < .086$). This discrepancy is likely due to the small sample size of the data set.

**Table 12.9   Omnibus Test for Multiple Regression**

|  |  |  |  | Model Summary |  |  |  |  |  |
|---|---|---|---|---|---|---|---|---|---|
|  |  |  |  |  |  | Change Statistics |  |  |  |
| Model | R | R Square | Adjusted R Square | Std. Error of the Estimate | R Square Change | F Change | df1 | df2 | Sig. F Change |
| 1 | .328[a] | .107 | .087 | .25150 | .107 | 5.178 | 1 | 43 | .028 |
| 2 | .332[b] | .110 | .068 | .25409 | .003 | .129 | 1 | 42 | .722 |

a. Predictors: (Constant), mhskills.
b. Predictors: (Constant), mhskills, studtch.

| | ANOVA[c] | | | | | |
|---|---|---|---|---|---|---|
| Model | | Sum of Squares | df | Mean Square | F | Sig. |
| 1 | Regression | .327 | 1 | .327 | 5.178 | .028[a] |
| | Residual | 2.720 | 43 | .063 | | |
| | Total | 3.047 | 44 | | | |
| 2 | Regression | .336 | 2 | .168 | 2.601 | .086[b] |
| | Residual | 2.712 | 42 | .065 | | |
| | Total | 3.047 | 44 | | | |

a. Predictors: (Constant), mhskills.
b. Predictors: (Constant), mhskills, studtch.
c. Dependent Variable: mhimpact.

2. What is the MLR equation for the final model?

Answer:

$$\hat{Y}_{(mhimpact)} = (.321)\, X_{(mhskills)} + (-.003)\, X_{(studtch)} + (1.799)$$

3. How does the $R^2$ value change as new predictors are added?

Answer: The $R^2$ for model 1 is .107 (adjusted $R^2$ = .087). When the second predictor is added, the $R^2$ increases by .003 (to .11). (The adjusted $R^2$ actually decreases, in large part due to the small sample size.) You can see this in the "R Square Change" column under the "Model Summary" panel. Adding the second predictor did not produce a significant increase, as noted under the "Sig. F Change" value of .722.

4. Examine the regression coefficients to determine which are statistically significant. Are the $t$ tests for individual predictors significant?

Answer: The $t$ test for mhskills is significant in both models as shown in the "Coefficients" panel of Table 12.10 below. In model 1, the $t$ test = 2.275 ($p < .028$), and in model 2 it is 2.259 ($p < .029$) The second predictor produces a $t$ test of ($-.359$) when it is added to the model, not a significant finding ($p < .722$). Notice that if you square the $t$ test values, you will see the results in the "Model Summary" panel of Table 12.9. The squared $t$ test for model 1, 5.18 ($2.275^2$), is represented in model 1's "F Change" column. Since there is only one predictor in the model, the overall $F$ value changes from 0 to reflect the impact of the single predictor. The second squared predictor $t$ test, .1288 ($-.359^2$), is found in the "F Change" column of model 2.

**Table 12.10   T-Tests for Individual Predictors**

| | | | Coefficients[a] | | | | | | |
| | Unstandardized Coefficients | | Standardized Coefficients | | | Correlations | | |
| Model | B | Std. Error | Beta | t | Sig. | Zero-order | Partial | Part |
|---|---|---|---|---|---|---|---|---|
| 1  (Constant) | 1.803 | .417 | | 4.319 | .000 | | | |
| mhskills | .306 | .134 | .328 | 2.275 | .028 | .328 | .328 | .328 |
| 2  (Constant) | 1.799 | .422 | | 4.264 | .000 | | | |
| mhskills | .321 | .142 | .344 | 2.259 | .029 | .328 | .329 | .329 |
| studtch | −.003 | .009 | −.055 | −.359 | .722 | .046 | −.055 | −.052 |

a. Dependent Variable: mhimpact.

5. Square the part correlations on the final model when all variables are entered to determine the effect sizes. Which effect size is largest?

Answer: The squared part correlations for the predictors in model 2 are

$$Mhskills = .1082\,(.329^2)$$

$$Studtch = .0027\,(-.052^2)$$

The effect size for mhskills is largest. Therefore, in this sample of data, mhskills accounts for a great deal more of the variance in mhimpact than studtch ratios.

Notice that the squared part correlation values can be located in the "Model Summary" of Table 12.9 under the "R Square Change" column. When additional predictors are added to the model, the added predictor's squared part correlation is registered under this column. Thus in model 1, because there is only one predictor, the $.328^2$ part correlation for mhskills is found under model 1 in the "R Square Change" column (.1076). It is also equivalent to the overall $R^2$ of the model with one predictor. When the second predictor is added, the squared part correlation is found in model 2 of the "R Square Change" column, since it is the last variable added to the model.

## REAL WORLD LAB—DUMMY CODING

The SPSS® data file showing the subvariables created for the trtmt-groups is shown in Figure 12.6.

1. What are the omnibus and individual predictor results?

Answer: The SPSS® results are shown in Table 12.11. According to the "Model Summary" panel, the omnibus ($R^2$) for the model is .406 (.355 adjusted $R^2$), which indicates that about 41% (36% adjusted) of the variance in classrating is accounted for by the three treatment groups. The ANOVA table indicates a statistically significant finding ($F = 7.867$, $p < .002$). As with other MLR analyses, the explained variance can be calculated from the sums of squares shown in the ANOVA table (.406 = 8868.26/21831.26).

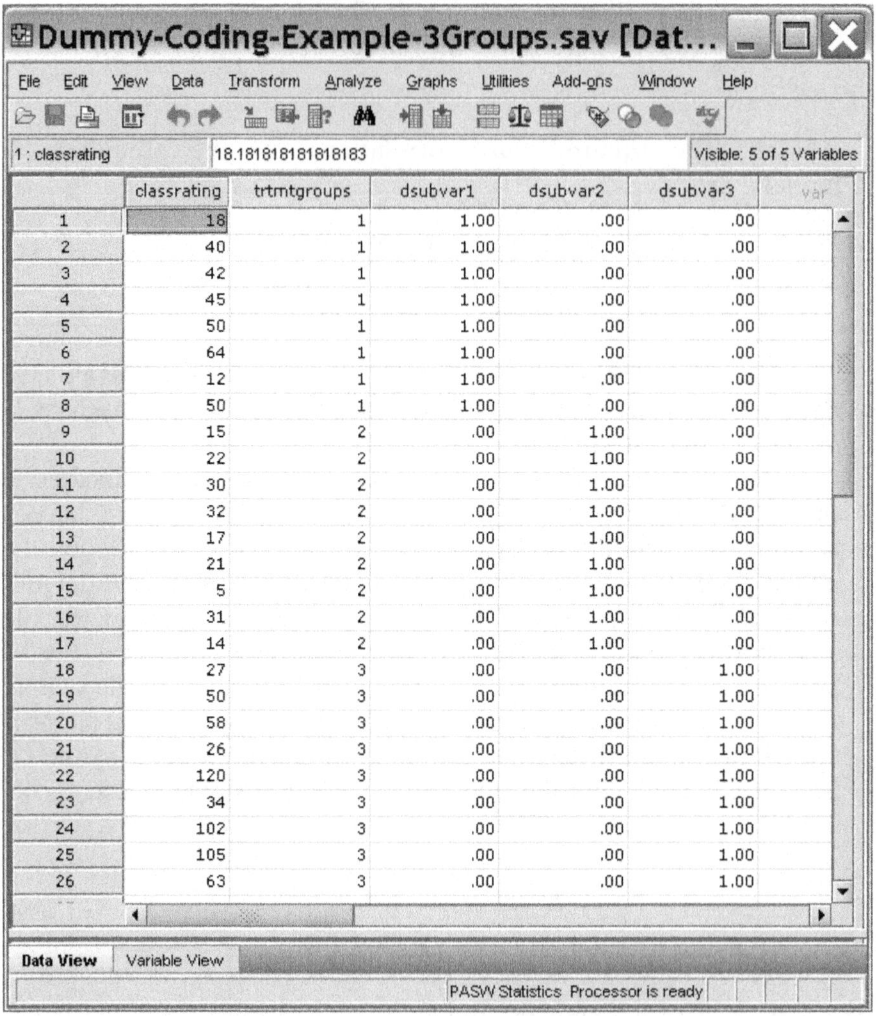


| | classrating | trtmtgroups | dsubvar1 | dsubvar2 | dsubvar3 | var |
|---|---|---|---|---|---|---|
| 1 | 18 | 1 | 1.00 | .00 | .00 | |
| 2 | 40 | 1 | 1.00 | .00 | .00 | |
| 3 | 42 | 1 | 1.00 | .00 | .00 | |
| 4 | 45 | 1 | 1.00 | .00 | .00 | |
| 5 | 50 | 1 | 1.00 | .00 | .00 | |
| 6 | 64 | 1 | 1.00 | .00 | .00 | |
| 7 | 12 | 1 | 1.00 | .00 | .00 | |
| 8 | 50 | 1 | 1.00 | .00 | .00 | |
| 9 | 15 | 2 | .00 | 1.00 | .00 | |
| 10 | 22 | 2 | .00 | 1.00 | .00 | |
| 11 | 30 | 2 | .00 | 1.00 | .00 | |
| 12 | 32 | 2 | .00 | 1.00 | .00 | |
| 13 | 17 | 2 | .00 | 1.00 | .00 | |
| 14 | 21 | 2 | .00 | 1.00 | .00 | |
| 15 | 5 | 2 | .00 | 1.00 | .00 | |
| 16 | 31 | 2 | .00 | 1.00 | .00 | |
| 17 | 14 | 2 | .00 | 1.00 | .00 | |
| 18 | 27 | 3 | .00 | .00 | 1.00 | |
| 19 | 50 | 3 | .00 | .00 | 1.00 | |
| 20 | 58 | 3 | .00 | .00 | 1.00 | |
| 21 | 26 | 3 | .00 | .00 | 1.00 | |
| 22 | 120 | 3 | .00 | .00 | 1.00 | |
| 23 | 34 | 3 | .00 | .00 | 1.00 | |
| 24 | 102 | 3 | .00 | .00 | 1.00 | |
| 25 | 105 | 3 | .00 | .00 | 1.00 | |
| 26 | 63 | 3 | .00 | .00 | 1.00 | |

**Figure 12.6** SPSS® screenshot showing dummy coding of classrating.

The individual predictors' results shown in Table 12.12 indicate that treatment 1 and treatment 2 are both different from treatment 3. Note that the coefficients are shown only for treatment subvariables 1 and 2 because those were the only subvariables included in the equation. Had we included the third subvariable, we would have created an untenable multicollinearity condition. We can examine the coefficients to understand the differences among the treatment groups.

**Table 12.11 Omnibus Results for Dummy Coding Example**

|  | | | | Model Summary | | | | | |
| --- | --- | --- | --- | --- | --- | --- | --- | --- | --- |
|  | | | | | | Change Statistics | | | |
| Model | R | R Square | Adjusted R Square | Std. Error of the Estimate | R Square Change | F Change | df1 | df2 | Sig. F Change |
| 1 | .637[a] | .406 | .355 | 23.740 | .406 | 7.867 | 2 | 23 | .002 |

a. Predictors: (Constant), dsubvar2, dsubvar1.

| | | ANOVA[b] | | | | |
| --- | --- | --- | --- | --- | --- | --- |
| Model | | Sum of Squares | df | Mean Square | F | Sig. |
| 1 | Regression | 8868.256 | 2 | 4434.128 | 7.867 | .002[a] |
| | Residual | 12963.004 | 23 | 563.609 | | |
| | Total | 21831.260 | 25 | | | |

a. Predictors: (Constant), dsubvar2, dsubvar1.
b. Dependent Variable: classrating.

**Table 12.12 Individual Results for Dummy Coding Example**

| | | Coefficients[a] | | | | | | | |
| --- | --- | --- | --- | --- | --- | --- | --- | --- | --- |
| | | Unstandardized Coefficients | | Standardized Coefficients | | | Correlations | | |
| Model | | B | Std. Error | Beta | t | Sig. | Zero-order | Partial | Part |
| 1 | (Constant) | 65.083 | 7.913 | | 8.224 | .000 | | | |
| | dsubvar1 | −24.999 | 11.536 | −.398 | −2.167 | .041 | −.046 | −.412 | −.348 |
| | dsubvar2 | −44.279 | 11.191 | −.727 | −3.957 | .001 | −.534 | −.636 | −.636 |

a. Dependent Variable: classrating.

Overall, the adapted results in this example show that in this sample, the constructivist teaching approach outperformed the traditional and ethnomath approaches on a math achievement test.

2. How do you interpret the constant and unstandardized B values?

Answer: The mean values by treatment groups in Table 12.13 will help show the values of the coefficients:

**Table 12.13   Mean Values of Treatment Groups**

| Treatment Groups | Classrating Means |
|---|---|
| 1 | 40.08 |
| 2 | 20.8 |
| 3 | 65.08 |

Recall that the constant value (65.083) is the mean of the subvariable group not in the equation. You can see that the treatment group 3 mean of classrating from the table (65.08) is equal to the constant value in the SPSS® output.

Using the tabled values, you can reconstruct the subvariable coefficients from the SPSS® output (Figure 12.6):

$$(-24.999) = 40.08 - 65.08 \qquad \text{treatment group 1}$$

$$(-44.279) = 20.80 - 65.08 \qquad \text{treatment group 2}$$

## REAL WORLD LAB—INTERACTION

In this exercise, you will run a MLR analysis in which there are categorical and continuous predictors and a product term. The hypothetical data involve school-level findings where the focus is on predicting achievement from a categorical and a continuous predictor. The variables in the database (at the end of Chapter 10) are as follows:

- Math achievement ("MathAch")—a continuous outcome
- Students per classroom ("Classratio")—a continuous predictor
- Family income ("FamilyIncome3")—a categorical predictor (3 groups) measuring the percentage of students at school with low, medium, and high family income.
- Dummy subvariables for FamilyIncome3: dsubvar1, dsubvar2, and dsubvar3.
- The product variable for the two predictors, which consists of three subvariables:
  ○ S1Clrat—Subvariable 1 × Classratio
  ○ S2Clrat—Subvariable 2 × Classratio
  ○ S3Clrat—Subvariable 3 × Classratio

1. Conduct a MLR analysis with a continuous predictor (Classratio) and a categorical (FamilyIncome3) predictor.

Answer: You will make this analysis in stages to see how to move from the overall MLR to the inclusion of product terms and eventually to graphing the interaction. Notice that I created subvariables for the categorical variable FamilyIncome3 and product subvariables in the database. I used this same procedure in the Interaction chapter, except that this database uses a three-group categorical variable with a continuous predictor.

It is first important to look at the overall MLR without the product term to see what the relationships are among the predictors and outcome. In order to run the overall MLR, enter the continuous variable first; next enter the two subvariables *as a set in a second block*. That way you can see the change in the MLR results when the subvariables are added (together) to the analysis that already contains the continuous variable (in the first block). See Table 12.14

The results of this analysis show the change in $R^2$ as a result of adding the two dummy subvariables:

**Table 12.14  Omnibus Output for Continuous and Categorical Predictors**

|  |  |  |  |  | Model Summary |  |  |  |  |
| --- | --- | --- | --- | --- | --- | --- | --- | --- | --- |
|  |  |  |  |  | Change Statistics | | | | |
| Model | R | R Square | Adjusted R Square | Std. Error of the Estimate | R Square Change | F Change | df1 | df2 | Sig. F Change |
| 1 | .169[a] | .029 | .005 | 21.748 | .029 | 1.206 | 1 | 41 | .279 |
| 2 | .375[b] | .141 | .075 | 20.971 | .112 | 2.547 | 2 | 39 | .091 |

a. Predictors: (Constant), Classratio.
b. Predictors: (Constant), Classratio, dsubvar1, dsubvar2.

These results show that the overall $R^2$ increased by .112 (to .141) as a result of adding the two subariables representing the dummy variable categories (dsubvar1 and dsubvar2). The first model shows that Classratio is not a significant predictor of MathAch (Sig. F Change = .279). The omnibus results for model 2 are likewise not significant (Sig. F Change = .091). However, looking at the coefficients in the second model, shown in Table 12.15, dsubvar1 (representing low family income) is significantly different from dsubvar3 (representing high income, which is not entered and therefore the comparison group).

**Table 12.15   Individual Predictor Output for Continuous and Categorical Predictors**

| | | Coefficients[a] | | | | |
|---|---|---|---|---|---|---|
| | | Unstandardized Coefficients | | Standardized Coefficients | | |
| Model | | B | Std. Error | Beta | t | Sig. |
| 1 | (Constant) | 45.333 | 9.597 | | 4.723 | .000 |
| | Classratio | −.555 | .506 | −.169 | −1.098 | .279 |
| 2 | (Constant) | 53.586 | 11.386 | | 4.706 | .000 |
| | Classratio | −.704 | .515 | −.214 | −1.368 | .179 |
| | dsubvar1 | −18.452 | 8.333 | −.362 | −2.214 | .033 |
| | dsubvar2 | −4.043 | 7.783 | −.088 | −.519 | .606 |

a. Dependent Variable: MathAch.

If, on the basis of these or other reasons, you suspect there may be an interaction, perform the interaction analysis:

- Create the product variable by multiplying the categorical subvariables and continuous variable ("clratXfamilyInc"). I included these in the database as I described above.
- Create the MLR analysis predicting MathAch, including interaction terms, by entering into the analysis (1) the continuous variable, then (2) the categorical subvariables as a set in their own block, and finally, (3) the product subvariables in their own block.

The resulting three-model  analysis is shown in Tables 12.16 and 12.17. The first panel (Model Summary) shows the effects on $R^2$ of adding the product subvariables. The $R^2$ increased by .211 as shown in the results for model 3. The change was significant, $p < .005$. This indicates that there is a significant interaction among the variables, so we need to examine the results carefully to interpret the relationships.

**Table 12.16   Effects of Subvariables on $R^2$**

<table>
<tr><td colspan="11" align="center">Model Summary</td></tr>
<tr><td></td><td></td><td></td><td></td><td></td><td colspan="6" align="center">Change Statistics</td></tr>
<tr><td>Model</td><td>R</td><td>R Square</td><td>Adjusted R Square</td><td>Std. Error of the Estimate</td><td>R Square Change</td><td>F Change</td><td>df1</td><td>df2</td><td>Sig. F Change</td></tr>
<tr><td>1</td><td>.169[a]</td><td>.029</td><td>.005</td><td>21.748</td><td>.029</td><td>1.206</td><td>1</td><td>41</td><td>.279</td></tr>
<tr><td>2</td><td>.375[b]</td><td>.141</td><td>.075</td><td>20.971</td><td>.112</td><td>2.547</td><td>2</td><td>39</td><td>.091</td></tr>
<tr><td>3</td><td>.593[c]</td><td>.352</td><td>.264</td><td>18.701</td><td>.211</td><td>6.021</td><td>2</td><td>37</td><td>.005</td></tr>
</table>

a. Predictors: (Constant), Classratio.
b. Predictors: (Constant), Classratio, dsubvar1, dsubvar2.
c. Predictors: (Constant), Classratio, dsubvar1, dsubvar2, S2xClrat, S1xClrat.

The next panel shows the results of analysis on the individual predictors. As the panel shows, S2xClrat is significant ($p < .001$), which may help explain the overall model results. You can also see that the "main effects" of class ratio and family income are significant.

**Table 12.17   Individual Predictor Analysis with Possible Interaction Effects**

<table>
<tr><td colspan="9" align="center">Coefficients[a]</td></tr>
<tr><td></td><td colspan="2" align="center">Unstandardized Coefficients</td><td align="center">Standardized Coefficients</td><td></td><td></td><td colspan="3" align="center">Correlations</td></tr>
<tr><td>Model</td><td>B</td><td>Std. Error</td><td>Beta</td><td>t</td><td>Sig.</td><td>Zero-order</td><td>Partial</td><td>Part</td></tr>
<tr><td>1  (Constant)</td><td>45.333</td><td>9.597</td><td></td><td>4.723</td><td>.000</td><td></td><td></td><td></td></tr>
<tr><td>   Classratio</td><td>−.555</td><td>.506</td><td>−.169</td><td>−1.098</td><td>.279</td><td>−.169</td><td>−.169</td><td>−.169</td></tr>
<tr><td>2  (Constant)</td><td>53.586</td><td>11.386</td><td></td><td>4.706</td><td>.000</td><td></td><td></td><td>—</td></tr>
<tr><td>   Classratio</td><td>−.704</td><td>.515</td><td>−.214</td><td>−1.368</td><td>.179</td><td>−.169</td><td>−.214</td><td>−.203</td></tr>
<tr><td>   dsubvar1</td><td>−18.452</td><td>8.333</td><td>−.362</td><td>−2.214</td><td>.033</td><td>−.315</td><td>−.334</td><td>−.329</td></tr>
<tr><td>   dsubvar2</td><td>−4.043</td><td>7.783</td><td>−.088</td><td>−.519</td><td>.606</td><td>.108</td><td>−.083</td><td>−.077</td></tr>
<tr><td>3  (Constant)</td><td>64.830</td><td>10.684</td><td></td><td>6.068</td><td>.000</td><td></td><td></td><td></td></tr>
<tr><td>   Classratio</td><td>−1.264</td><td>.488</td><td>−.385</td><td>−2.591</td><td>.014</td><td>−.169</td><td>−.392</td><td>−.343</td></tr>
<tr><td>   dsubvar1</td><td>−48.878</td><td>74.565</td><td>−.958</td><td>−.656</td><td>.516</td><td>−.315</td><td>−.107</td><td>−.087</td></tr>
<tr><td>   dsubvar2</td><td>−86.252</td><td>24.736</td><td>−1.876</td><td>−3.487</td><td>.001</td><td>.108</td><td>−.497</td><td>−.462</td></tr>
<tr><td>   S1xClrat</td><td>1.682</td><td>4.329</td><td>.566</td><td>.389</td><td>.700</td><td>−.313</td><td>.064</td><td>.051</td></tr>
<tr><td>   S2xClrat</td><td>5.203</td><td>1.504</td><td>1.796</td><td>3.460</td><td>.001</td><td>.203</td><td>.494</td><td>.458</td></tr>
</table>

a. Dependent Variable: MathAch.

Having determined that there is a significant interaction, we can proceed as we did in Chapter 10 to examine the results graphically. In order to see how the interaction looks graphically:

- Use "Split File" in SPSS® specifying the categorical variable as the grouping variable (FamilyIncome3). This will create three output analyses, one for each treatment group (1-low, 2-medium, and 3-high family income).

- Run the MLR predicting MathAch from the continuous variable (classratio). You do not need to specify the categorical or product variables in this analysis, since they are "contained" in the split file process.

- Your output file should contain three separate analyses, one for each treatment condition as shown in Tables 12.18, 12.19 and 12.20. You can use the resulting regression equations to create a table to generate values to use for creating the graph.

**Table 12.18   FamilyIncome3—Results for Group "1" Low Income**

| | | | | | | | | | |
|---|---|---|---|---|---|---|---|---|---|
| | | | | Coefficients[a,b] | | | | | |
| | | Unstandardized Coefficients | | Standardized Coefficients | | | | Correlations | |
| Model | | B | Std. Error | Beta | t | Sig. | Zero-order | Partial | Part |
| 1 | (Constant) | 15.952 | 44.286 | | .360 | .728 | | | |
| | Classratio | .418 | 2.581 | .057 | .162 | .875 | .057 | .057 | .057 |

a. FamilyIncome3 = 1.00.
b. Dependent Variable: MathAch.

**Table 12.19   FamilyIncome3—Results for Group "2" Medium**

| | | | | | | | | | |
|---|---|---|---|---|---|---|---|---|---|
| | | | | Coefficients[a,b] | | | | | |
| | | Unstandardized Coefficients | | Standardized Coefficients | | | | Correlations | |
| Model | | B | Std. Error | Beta | t | Sig. | Zero-order | Partial | Part |
| 1 | (Constant) | −21.422 | 19.281 | | −1.111 | .288 | | | |
| | Classratio | 3.939 | 1.229 | .679 | 3.204 | .008 | .679 | .679 | .679 |

a. FamilyIncome3 = 2.00.
b. Dependent Variable: MathAch.

**Table 12.20    FamilyIncome3—Results for Group "3" High**

Coefficients[a,b]

| Model | | Unstandardized Coefficients | | Standardized Coefficients | | | Correlations | | |
|---|---|---|---|---|---|---|---|---|---|
| | | B | Std. Error | Beta | t | Sig. | Zero-order | Partial | Part |
| 1 | (Constant) | 64.830 | 12.996 | | 4.988 | .000 | | | |
| | Classratio | −1.264 | .594 | −.459 | −2.130 | .048 | −.459 | −.459 | −.459 |

a. FamilyIncome3 = 3.00.
b. Dependent Variable: MathAch.

Once you have the split file results, you can create a table to generate values to graph. You will do this by including the regression coefficients for each of the split file results, and predicting values of the outcome variable using high, medium, and low values of the continuous predictor. In this example I used the mean (17.81) plus/minus one standard deviation (6.84) of Classratio to generate low, medium, and high values (11.17, 17.81, 24.45).

Table 12.21 shows all the values needed to generate the graph (I used Microsoft Excel to generate the line graph). The top part of the table reproduces the regression equations for each group of the categorical variables.

The middle section of the table shows the values of the continuous predictor (Classratio) that I used with the separate equations to predict values of math achievement for each group.

The bottom part of the table shows the results of using the low, medium, and high values of Classratio in the equations. For example, the top left value in the shaded cell represents the predicted math achievement value using 11.17 in the equation for FamilyIncome1.

**Table 12.21    Table of Values for Interaction Graph**

| | | FamilyIncome1 | FamilyIncome2 | FamilyIncome3 |
|---|---|---|---|---|
| Intercept | | 15.95238095 | −21.42231405 | 64.83009674 |
| b value | | 0.417989418 | 3.938842975 | −1.264493013 |
| Low/Med/High values of Classratio | | | | |
| Low | 11.17 | | | |
| Medium | 17.81 | | | |
| High | 24.45 | | | |
| Y' using Classratio Low | | 20.62132275 | 22.57456198 | 50.70570978 |
| Y' using Classratio Med | | 23.39677249 | 48.72847934 | 42.30947617 |
| Y' using Classratio High | | 26.17222222 | 74.88239669 | 33.91324257 |

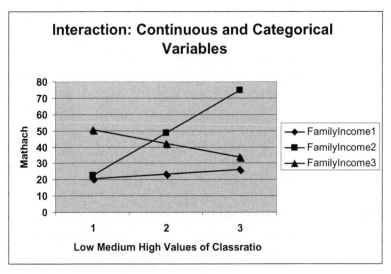

**Figure 12.7** Interaction of continuous and categorical predictors.

You can use the (Line) graph function in Microsoft Excel to generate an interaction graph using the nine cells at the bottom of Table 12.21. The resulting graph (Figure 12.7) shows the interaction among the three groups of family income.

You can see the interaction through the differential directions of the separate lines. The classic interaction indication is a crossing of the separate group lines, in this case, between medium and high family incomes. As the graph shows, when family income is low, there appears to be little impact of class size on math achievement. However, when the family income increases, there are differential impacts on the relationship between class size and math achievement. At medium family incomes, math achievement increases along with class size. At high incomes, math achievement declines as class sizes increase.

# APPENDIX

**Washington School Research Center**

Technical Report #1 – July 2001

# The Relationships Among Achievement, Low Income, and Ethnicity Across Six Groups of Washington State Students

## Martin L. Abbott, Ph.D.

## Jeff Joireman, Ph.D.

**The Washington School Research Center (WSRC)** is an independent research and data analysis center within Seattle Pacific University. The Center began in July 2000, funded through a gift from the Bill and Melinda Gates Foundation. Our mission is to conduct sound and objective research on student learning in the public schools, and to make the research findings available for educators, policy makers, and the general public for use in the improvement of schools. We believe that sound data and appropriate data analysis are vital components for the identification of school and classroom practices related to increased student academic achievement.

Washington School Research Center
3500 188<sup>th</sup> St. S.W., Suite 328
Lynnwood, WA 98037
Phone: 425.744.0992
Fax: 425.744-1241
Web: www.spu.edu/wsrc

| **Jeffrey T. Fouts** | **Martin L. Abbott, Ph.D.** | **Duane B. Baker, Ed.D.** |
|---|---|---|
| Executive Director | Senior Researcher | Director - |
| Professor of Education | Professor of Sociology | School Information Services |

# The Relationships Among Achievement, Low Income, and Ethnicity

## Across Six Groups of Washington State Students

A Technical Report For
The Washington School Research Center

Washington School Research Center

# Foreword

We are pleased to provide this technical report on the relationships among student achievement, income, and ethnicity as the first publication for Washington educators by the Washington School Research Center. The questions addressed by these analyses are important considerations for all of us concerned with improving education in the state of Washington. Media reports that highlight the different achievement levels of various ethnic groups of children are common. These differences are a source of great concern among community groups, and rightfully so. Yet, those of us who work with data and statistics on a regular basis are acutely aware of the dangers inherent in reporting group achievement results that consider only one characteristic for creating those groups.

Factors affecting student achievement are varied and complex, and failure to consider multiple factors may lead to erroneous or simplistic answers to very complicated questions. In this report, professors Abbott and Joireman address the question of differences in school level achievement depending on the ethnic composition of the student population, so often reported in the media, while at the same time considering the income levels of the students' families. They begin this effort with a brief review of research conducted elsewhere on this topic, and conclude that previous research has shown that "income is generally a better predictor of student achievement than ethnicity."

Using aggregate school 3$^{rd}$ & 6$^{th}$ grade ITBS test scores for 1999 and 2000, 4$^{th}$ grade WASL scores for 1999 and 2000, and 7$^{th}$ grade WASL scores for 1999 for all schools in the state, Abbott and Joireman examine the relationships among these scores and the percentage of students receiving free or reduced lunch at the school, and the various percentages of students comprising a variety of ethnic groups. Using a statistical procedure called multiple regression, they are able to determine the relative importance of these latter two variables in determining the schools' achievement levels. Their findings? "Across a variety of grades and tests, our results support the conclusion that low income explains a much larger percentage of the variance in academic achievement than ethnicity."

Abbott and Joireman do not say that ethnicity is unimportant or unrelated to achievement, but low income appears to be a much more influential factor. They conclude that, "the relationship between ethnicity and academic achievement is mostly indirect: ethnicity relates to low income and low income relates to academic achievement . . ." In other words, low income is the stronger predictor of school achievement, and non-white families are over-represented among the low incomes. These findings suggest therefore, that schools with predominately white, low income populations have achievement levels more in common with schools with non-white, low income populations than they do with schools with white, high income populations. Conversely, the achievement levels of schools with high income student populations more closely resemble other schools with high income student bodies irrespective of their ethnic composition.

Educators throughout the state, indeed throughout the country, are striving to raise the achievement levels of all students. A student's ethnicity is often an observable student characteristic that is frequently viewed as a determinant of that student's achievement level. However, these and other results suggest that it is the effects of poverty that play a much larger role in a student's chance for success in school, and it is those effects that educators and policy makers should consider first as prevention, intervention, and remedial programs are designed.

Jeffrey T. Fouts
Executive Director
Washington School Research Center
Lynnwood, WA

# Table of Contents

Introduction ............................................................................ 2

Method ................................................................................... 6

Results .................................................................................. 8

Discussion ............................................................................16

Literature Cited ..................................................................... 19

Table 1 ................................................................................. 21

Table 2 ................................................................................. 22

Table 3 ................................................................................. 23

Table 4 ................................................................................. 24

# INTRODUCTION

Educators at all levels face challenges to learning that are embedded in the student's background and characteristics, as well as systemic to the particular learning program and leadership system in place in the school. While many of these may never be fully understood, it is important to the success of the educational effort to acknowledge that all students can learn under the proper circumstances. Part of the attempt by educational research programs should be to identify and understand the complex nature of the learning environment, and what salient factors are likely to lead to successful learning experiences.

This technical report is an attempt to do just that. By looking in detail at a set of Washington school data, the report is designed to respond to questions from those who are concerned about student performance, and who have asked specifically about the potential impact of certain variables on student learning. One of these questions concerns the interrelationship of low-income and ethnicity on student achievement. That is, are both low-income and ethnicity equally likely to impact student learning? Or, are the two so intertwined that it is difficult to understand the specific influence of each on student achievement?

In the following pages, we attempt to respond to these questions by reporting on data analyses designed to identify the unique effects of predictor variables on student achievement. These analyses are not intended to be exhaustive or definitive with respect to these questions, but rather to add to an existing inquiry by other researchers. The body of existing research will be examined for trends that might be supported or contradicted by the data analyses in this Technical Report.

It is important at the outset to point out what a technical report is not. It is not designed to be a comprehensive analysis of all possible determinants of student achievement. Cataloging the entire range of influence on achievement would be an unrealistic goal of any research program. Additionally, a technical report oftentimes cannot hope to generalize the findings of research beyond the database used in the analysis due to the nature of the limitations of the database itself (e.g., using data that have already been defined and collected, or that may not perfectly operationalize a research construct).

Because the focus of the technical report is on the technical detail of the data, there is no attempt to use the information gained to drive or change policy recommendations. Rather, the attempt is simply to report what is discovered in the hopes that the findings will help to clarify factors non-conducive to learning, and those that might be useful in broader efforts to improve learning. No claims are made that the analyses will resolve the longstanding, and oftentimes highly charged, debates about the roles of ethnicity and poverty in learning.

Two large-scale studies point to the importance of studying further the relationship between ethnicity and low income. As part of Jencks and Phillips', The Black-White Test Score Gap (1998), Hedges and Nowell suggest that socioeconomic factors affect the Black-White gap in test scores, but cannot entirely explain the "black-white test score convergence" (p. 167). While the current technical report does not address the Black-White test score gap specifically, it does focus on what possible impact ethnicity might have on achievement, taking income into account. In this way, it may contribute to a broader understanding of the dynamics that affect all ethnic groups.

Reaching the Top, the recently published Report of the National Task Force on Minority High Achievement (1999), is an excellent attempt to summarize the thinking and dialog on the overall issue of minority achievement. While reactions to this document are diverse (for example, see the symposium reactions in Society, July/August, 2000), and are as often filled with invective as with praise, it is apparent that further research-based efforts are needed to address the complexity of the ethnicity-poverty relationship. Hopefully, this technical report will be a contribution to that end.

## Related Research Literature

Taken together, the recent research literature examining the unique effects of low income and ethnicity on student achievement is not conclusive. In many cases, the statistical analyses are not pointed toward disentangling the separate effects of the variables, and/or the overall research problems encompass other targets. However, the few carefully conducted studies we highlight suggest that, when studied together, low income is generally a better predictor of student achievement than ethnicity.

In some cases, low income and/or ethnicity may appear to be meaningfully related to student achievement when examined individually (Wong and Alkins, 1999; Yellin and Koetting, 1991; Fenwick, 1996). This may be due to the fact that poverty and ethnicity are often coterminous. That is, students of some ethnic backgrounds also may be those who are unequally represented in low-income families. However, the specific contribution of each of these to an understanding of achievement may be confounded by their interrelationship to one another (Patterson, et. al., 1990). For this reason, it is important to move beyond univariate analyses, or more general attempts, and examine the contribution of each predictor variable with student achievement when the other predictor is present in the analysis. (Dulaney and Banks' 1994 descriptive study is a helpful step in this respect.)

Some insight into this matter is found through studies that attempt to understand the relationship between student achievement and independent factors other than income and ethnicity. Desimone's (1999) study, for example, concluded that there were statistically significant and meaningful differences between parent involvement and achievement, according to race-ethnicity and family income. However, the unique contributions of race-ethnicity and family income upon achievement were not fully elaborated. This was also the case in Johnson's (2000) study of peer effects, and reports examining student mobility (Bolinger and Gilman, 1997) and curriculum alignment (Mitchell, 1999).

More compelling evidence of the specific impact of ethnicity and income on achievement is provided by studies that are statistically tailored and/or expressly focused on the interrelationships of these key variables. Most all of these indicate that income provides the greatest impact on student achievement when the effects of ethnicity are taken into account. Peng and Wright's (1994) analysis of academic achievement, home environments (including family income), educational activities, and ethnicity, concluded that when all variables were included in the analysis, ethnicity accounted for only a very small proportion of the variance (3%) in student achievement. Home environment and educational activities explained the greatest amount of variance (although the specific impact of income was not disaggregated from this group of variables).

The study by Patterson, et. al. (1990) cited earlier, provides another carefully controlled analysis of poverty and ethnicity (along with gender and household composition) as they relate to school-based competence (i.e., conduct, peer relations, and academic achievement). While ethnicity was cited as a strong predictor of achievement, income level and gender emerged as stronger overall predictors of the dependent variables, with income level the strongest predictor of achievement.

Two additional studies point to the importance of income in explaining achievement relative to ethnicity. Miller-Whitehead's (1999) study using hierarchical regression concluded that free/

reduced lunch status accounted for the majority of unexplained variance in science scale scores across grades 3 through 8. With grade 5 data, class size accounted for the greatest amount of the explained variability of science scores, while ethnicity was considered "marginally significant" (p.16), with free/reduced lunch status having little direct effect. The author concluded that the latter findings might be explained by the programs put in place in Tennessee schools to improve science achievement in fifth grade among low-income students. In examining alternative assessment methods in elementary science, Saturnelli and Repa (1995) conclude, with respect to the question of whether race or economic status has the greater effect on science and math achievement, "based on the results of this study, it appears that for science, the answer is economic status. Within each racial group, test scores were found to increase significantly from high-poverty to no-poverty levels (p.34)."

# METHOD

## Aggregation and Selection of Schools

The analyses presented in this report are based on aggregated, 1999 and 2000 school-level data obtained from the Washington State Office of the Superintendent of Public Instruction. Combining individual student responses within schools was necessary, as information about low income (i.e., percentage of students on free lunch) was only available for each school. Schools with less than ten students were excluded from the analyses since such cases would provide a less credible basis for a stable set of results. To examine the stability and generalizability of the findings, we examined the relationships among low income, ethnicity, and academic achievement within six groups (3 grade levels × 2 achievement tests).

## Measures of Low income, Ethnicity, and Academic Achievement

Consistent with past research, <u>low income</u> was defined as the percentage of students in a given school who were on free or reduced lunch. While the percentage of students on free/reduced lunch is not a direct measure of low income, it is at present the best existing measure, and it is used extensively throughout comparable research literature.

Across the six groups, six categories of ethnicity were identified, including Native/American Indian, Asian American, African American/Black, Hispanic, White, and Multi-Racial. For the purposes of our analyses, <u>ethnicity</u> was defined as the percentage of White students in a school. Obviously, this index does not allow for an evaluation of how different distributions of minority groups may relate to achievement. To be sure we were not overlooking potentially important information about such differences, we conducted a series of preliminary analyses incorporating the percentage of students in additional ethnic categories. Results from these analyses indicated that the inclusion of additional ethnic categories did not aid in the prediction of achievement.[1] As a result, analyses of additional ethnic categories are not included in our report.

---

[1] Preliminary analyses were run on the WASL only. Separate analyses were run for each additional ethnic category, adding the category in question, over and above low income and percentage of white students. With the exception of the percentage of Asian Americans, the additional ethnic categories failed to reach statistical significance. The percentage of Asian Americans showed a significant positive relationship with achievement, over and above low income and percent white. While statistically significant, the addition of this variable (% of Asian students) explained a relatively small percentage of the variance in achievement, ranging from 1.2% for WASL-Listening to 5.2% for WASL-Writing. As such, the addition of this variable would have a negligible practical impact on the findings presented in this report.

---

Academic achievement was assessed by using two statewide tests including the Washington Assessment of Student Learning (WASL) and the Iowa Test of Basic Skills (ITBS). Both tests contain scales assessing learning within four general domains. The WASL's four general domains include reading, math, listening, and writing. The ITBS's four general domains include reading, math, language (spelling, punctuation), and vocabulary.[2] Within these domains, both tests also contain more narrowly defined subscales. The present report focuses on the relationships among low income, ethnicity, and achievement in the broader learning domains. Future reports could focus on the relationships among low income, ethnicity, and achievement within the more narrowly defined domains.

## RESULTS

### Group Characteristics

Key characteristics of the six groups, including number of districts and schools, distribution of ethnic groups, as well as means and standard deviations for the various achievement tests are summarized in Tables 1 (WASL) and 2 (ITBS).

Relationship of Low income and Ethnicity to Achievement: Theoretical Models and Data Analytic Strategy

As noted in the introduction, the primary goal of the current report is to determine the relative contribution of low income and ethnicity to academic achievement. To evaluate the relationships among low income, ethnicity, and academic achievement, we conducted a series of multiple regression analyses. To set up the logic of these analyses, we begin by

---

[2]For more information on administration of the WASL and ITBS in Washington, visit www.k12.wa.us/assessment/WASLintro.asp. For more technical information on the WASL, visit www.k12.wa.us/assessment/qawasl.asp. For more technical information on the ITBS, visit www.riverpub.com/products/group/itbs.htm.

discussing three possible models, which may explain the relationships among ethnicity, low income, and academic achievement. These models, shown below, serve as a guide to the analyses included in this report.

Model 1, shown below, assumes that the two-predictor variables, ethnicity and low income, are each related to academic achievement directly. If true, the simple correlation of ethnicity and low income, respectively, with academic achievement should not change dramatically when the other predictor variable is taken into account. That is, the predictor variables (ethnicity and low income) should each have a sizeable and unique relationship with academic achievement, once the other predictor variable has been statistically controlled.

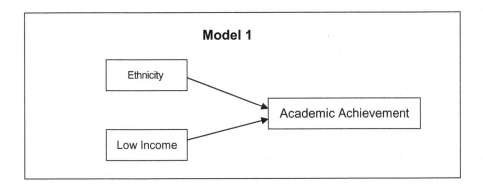

Model 2, shown below, assumes that ethnicity is related to academic achievement indirectly through its relationship to low income. Four conditions must be met for this model to receive support. First, ethnicity must be related to achievement (without low income in the model). Second, ethnicity must be related to low income. Third, low income must be related to academic achievement (without ethnicity in the model). Fourth, ethnicity's relationship with academic achievement should no longer be "significant," once the effect of low income on academic achievement has been statistically controlled.

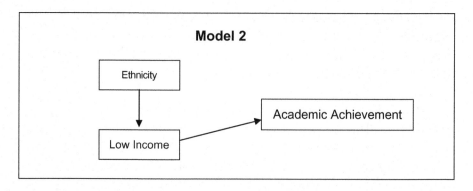

Model 3, shown below, assumes that ethnicity is related to academic achievement both indirectly, through its relationship to low income (as in Model 2), and directly, over and above its effect on low income (as in Model 1). Four conditions must be met for this model to receive support. First, ethnicity must be related to achievement (without low income in the model). Second, ethnicity must be related to low income. Third, low income must be related to academic achievement (without ethnicity in the model). Fourth, ethnicity's relationship with academic achievement should remain "significant," once the effect of low income on academic achievement has been statistically controlled.

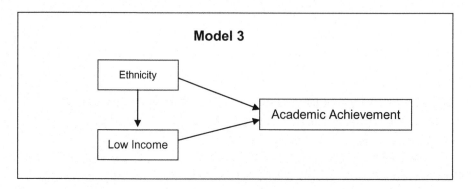

Model 2 vs. Model 3: The only difference between Models 2 and 3 is that Model 2 assumes that ethnicity is not a significant predictor of achievement, over and above the effect of low income, whereas Model 3 requires ethnicity to be a "significant" predictor of academic achievement, over and above the effect of low income. Choosing between Models 2 and 3 will therefore depend on how the term "significant" predictor is defined. A complete picture of whether a result is "significant" requires an evaluation of both its statistical and practical significance.

Statistical significance refers to the probability that any given result is due to chance. Traditionally, if this probability is less than 5%, researchers conclude that the result is unlikely to have occurred by chance, and consequently say that the result is "statistically significant." While statistical significance is an important benchmark in evaluating whether a result is likely to be due to chance, it can also be misleading if it is used as the only basis for determining whether a result is "significant" in the broader sense. One of the biggest problems with statistical significance is that it is heavily influenced by sample size. All things being equal, larger sample sizes will produce more statistically significant findings. In the case of very large sample sizes, even relatively small relationships will be "statistically significant." Because most of our analyses are based on more than 1000 schools, there is a good chance that even very small relationships will be statistically significant. As such, we will place more emphasis on the practical significance of the findings.

Practical significance refers to the size of any given result. Given the nature of our data, the size of any given result can be gauged in terms of the percentage of variance in academic achievement explained by low income and ethnicity, respectively. As discussed below, our analyses reveal that, overall, low income and ethnicity together typically explain between 40% and 60% of the variance in academic achievement, depending on the grade and achievement test in question. While this is useful information, our central goal is to understand how much of the overall variance in academic achievement is uniquely due to ethnicity and low income, respectively. Returning to the Models 2 and 3, the central question is whether ethnicity uniquely explains a practically significant amount of the variance in academic achievement, over and above the influence of low income. We address this question, and evaluate Models 1-3, in the next section of this report.

## Multiple Regression Analyses: Summary of Findings

To aid in the discussion of our findings, we present a brief summary of our results before moving into the details of the various analyses. In terms of the three models outlined earlier, the results of our analyses are most consistent with a weak version of Model 3, as depicted in the diagram below. Of the two predictors, low income is clearly the strongest, uniquely explaining 12 to 29% of the variance in achievement, depending on the grade and test. By comparison, ethnicity uniquely explains a much smaller 0 to 6% of the variance in achievement, again depending on the grade and test. Additional analyses indicate that ethnicity explains, on average, 32.76% of the variance in low income (average correlation, $r = -.57$). In summary, our results suggest that the relationship between ethnicity and academic achievement is mostly indirect: ethnicity is related to low income, which in turn is related to academic achievement, though ethnicity does show a small direct relationship with academic achievement, over and above the effect of low income. Restated, of the two predictor variables, low income is the most closely related to academic achievement, irrespective of ethnicity. More detail regarding these findings is presented in the next section.

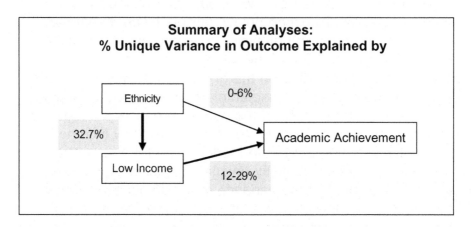

## Multiple Regression Analyses: Detailed Findings

To evaluate the relationships among low income, ethnicity, and academic achievement, we conducted a series of 24 multiple regression analyses (3 grades × 2 achievement tests × 4 achievement test subscales). Within each analysis, low income (% of students in a school on free/reduced lunch) and ethnicity (% of white students in a school) were entered as a set. Preliminary analyses indicated that the percentage of students in additional ethnic categories did not aid greatly in the prediction of achievement, over and above low income and the percentage of white students (see Footnote 1). As a result, the percentage of students in additional ethnic categories was not included in the analyses reported below. Results from the 24 multiple regression analyses are summarized in Tables 3 (WASL) and 4 (ITBS).

We begin a discussion of the results in the top left portion of Table 3. This section summarizes the relationships among low income (% students on free lunch), ethnicity (% of white students), and reading scores on the WASL for 4$^{th}$ grade tested in 1999 (WASL-4-99). Several aspects of these results deserve comment.

First, shown under the heading $R^2$ – Tot is the total variance in WASL-4-99 reading scores that is explained by low income and ethnicity as a set. These values have a possible range from 0 (0%) to 1 (100%). As can be seen in the table, over half of the variance in WASL-4-99 reading scores (55%) is explained by low income and ethnicity together as a set.

Second, shown under the heading R are the simple Pearson-Product Moment correlations between reading scores and each predictor variable. These values range from −1 (perfect negative relationship) to + 1 (perfect positive relationship). As can be seen, reading scores are negatively correlated with the percentage of students on free lunch in a given school ($r = -.72$), and positively correlated with the percentage of white students in a given school (r = .57). In other words, reading scores are lower in schools with a higher percentage of students on free lunch, and they are higher in schools with a higher percentage of white students.

Third, shown under the heading $R^2 - Ch$ is the percentage of variance in reading scores that is uniquely accounted for by each predictor variable (low income and ethnicity).[3] In theory, these values have a possible range from 0 (0% unique variance) to 1 (100% unique variance). This is arguably the most important part of the output, as it directly addresses the question of whether ethnicity predicts achievement, over and above the effect of low income. As can be seen in the table, the $R^2 - Ch$ values indicate that low income uniquely explains 23% of the variance in reading scores, while ethnicity uniquely explains only 3% of the variance in reading scores.[4] In other words, once low income is taken into account, ethnicity explains very little additional variance in reading scores. Based on these results, the most important predictor of reading scores is low income. Given the possible range for these values (0-100%), it may not appear very impressive that low income explains only 23% of the variance in reading scores. Indeed, that leaves 77% of the variance in reading scores unexplained. However, by many behavioral research standards, 23% is an impressive (practically significant) finding for a single predictor variable.

---

[3]R2 − Ch (i.e., $R^2$ change values) were computed by squaring the part correlation for each predictor variable. Identical values could have been obtained by conducting hierarchical multiple regression analyses in which we evaluated the percentage of variance each predictor added to the model, once the other predictor had been statistically controlled (i.e., $R^2$ change for the predictor entered into the model on the second step).

[4]It will be noted that the sum of these unique variances (26%) is less than the $R^2$-Tot (55%) discussed earlier. This is due to the fact that the unique variances do not take into account "shared variance", or variance in reading scores which is explained by low income and ethnicity, but which cannot be uniquely attributed to either variable.

Finally, shown under the column B are the unstandardized regression coefficients. Interested readers may use the unstandardized regression coefficients to predict WASL-4-99 reading scores by inserting relevant values for low income and ethnicity into a three-parameter regression model. For example, assume a school in question has 40% of its students on free/reduced lunch (B = .40), and is composed of 50% white students (B = .50). Using this information, the predicted WASL-4-99 reading score for that school would be 398.649: Predicted Reading Score = 406.27 + (−19.14 × .40) + (.07 × .50) = 406.27 + (−7.656) + (.035) = 398.649. This compares to the average WASL-4-99 reading score of 403.99 (shown in Table 1).

Having described in detail the various portions of the output for WASL-4-99 reading scores, we now proceed to a more general discussion of the results. Because reading and math scores are the most comparable subscales on the WASL and ITBS, we focus our discussion primarily on these two domains.

Comparing the results of the various regression analyses, several patterns seem worth mentioning. First, an examination of the $R^2$ − Tot values reveals that low income and ethnicity as a set tend to explain more of the variance in reading scores (average = 55.8%) than in math scores (average = 45.7%). Second, and more important, an examination of the $R^2$ − Ch values reveals that across all six groups, low income uniquely explains a much larger percentage of the variance in reading and math scores when compared to ethnicity. Averaging across grade levels and tests (WASL, ITBS), low income explains 24% of the variance in reading and 21.2% of the variance in math. By contrast, ethnicity explains only 3.5% of the variance in reading and only 2% of the variance in math. Reframed, low income explains 6.9 times more variance in reading, and 10.6 times more variance in math, when compared to ethnicity. Third, additional comparisons of the $R^2$ − Ch values on the WASL reading and math scores reveals a small trend for low income to become somewhat less important from 4th to 7th grade

(21.5% to 17%), and ethnicity to become somewhat more important from 4th to 7th grade (3% to 4.5%).

We now turn our attention to the remaining tests on the WASL (Listening and Writing) and ITBS (Language and Vocabulary). With regard to the WASL, and averaging over groups, it is apparent that, as a set, low income and ethnicity explain more of the overall variance ($R^2$ – Tot) in listening (47%) than in writing (36%). Focusing on the unique relationships ($R^2$ – Ch), it is apparent that low income explains relatively more of the variance in writing (19.5%) than in listening (15.3%), whereas ethnicity explains relatively more of the variance in listening (4.7%) than in writing (0.5%). Turning to the ITBS, and averaging over groups, it is apparent that, as a set, low income and ethnicity explain more of the overall variance in vocabulary (58.6%) than in language (32%). Similarly, low income and ethnicity each tend to explain more of the unique variance in vocabulary (25.3%, 4.3%) than in language (20.3%, 0.7%).

In summarizing these various findings, it is clear across a variety of grade levels, instruments (WASL, ITBS), and subscales on those instruments that low income explains the bulk of the variance in academic achievement (12-29%) when compared to ethnicity (0-6%). Additional analyses indicate that ethnicity explains over a third of the variance in low income (32.7%). Taken together, these results most strongly support a weak version of Model 3. That is to say, the relationship between ethnicity and academic achievement appears to be mostly indirect: ethnicity is related to low income, which in turn is related to academic achievement, though ethnicity does show a small direct relationship with academic achievement, over and above the effect of low income.

## DISCUSSION

The primary goal of this investigation was to evaluate the unique contribution of low income and ethnicity to academic achievement. Across a variety of grades and tests, our results support the conclusion that low income explains a much larger percentage of the variance in academic achievement than ethnicity. This is not to say that ethnicity is unrelated to academic achievement. Indeed, it is. The question is whether ethnicity influences academic achievement over and above the effects of low income. In response to that question, our results indicate that ethnicity explains between 0 to 6% of the variance in academic achievement, after the contribution of low income has been statistically controlled. Low income, by contrast, explains between 12 and 29% of the variance in academic achievement. Combined with the finding that ethnicity explains approximately 32.7% of the variance in low income, our results suggest that the relationship between ethnicity and academic achievement is mostly indirect: ethnicity relates to low income, and low income in turn relates to academic achievement (a weak version of Model 3, outlined earlier).

Before concluding, several limitations of the current report should be mentioned. First, because our analyses focused on aggregated, group-level data, the findings in this report, while suggestive, cannot be directly generalized to individual students.[5] Future research using student-level data could help clarify whether these results generalize to that level. Second, while low income and ethnicity together explained a relatively high percentage of the variance in most of the outcome measures, a sizable percentage of the variance in achievement scores could not be accounted for by these variables. For example, 44.2% of the variance in reading, and 54.3% of the variance in math, was unexplained by low income and ethnicity as a set. This clearly indicates that

---

[5]This technical report addresses aggregated (school level) data from the State of Washington. Analyses based on schools within individual districts generally reflect the overall study findings, but may show some slight discrepancies due to sample size or unique factors within the district. Subsequent exploration based on individual student-level data, when the data are available, will provide further insight into the impact of income and ethnicity on student achievement.

additional variables contribute to achievement within these domains. Future research should take these into account as a way of further explaining variations in student achievement.

Finally, the data presented here are correlational in nature. As with any correlational data, it is important to recognize that these data do not conclusively prove causation. While certain causal alternatives can be eliminated (e.g., academic achievement cannot influence ethnicity), there may be several different explanations for the relationships demonstrated here. In this regard, one important set of questions focuses on why low income is related to academic achievement (i.e., what mediates the relationship between low income and academic achievement?). To the extent that these mediating variables can be the target of interventions within or outside the schools, it may be possible to reduce the relationship between low income and academic achievement.

# REFERENCES

Bolinger, K., & Gilman, D. (1997). Student mobility and demographics: Relationships to aptitude and achievement in a three-year middle school. Terre Haute, Indiana. (ERIC Document ED409273)

The College Board. (1999). Reaching the top: A report of the national task force on minority high achievement. New York, NY: Author.

Desimone, L. (1999). Linking parent involvement with student achievement: Do race and income matter? The Journal of Educational Research, 93 (1), 11-30.

Dulaney, C., & Banks, K. (1994). Racial and gender gaps in academic achievement (E&R Report No. 94.10). Raleigh, NC: Wake County Public Schools System, Dept. of Evaluation and Research. (ERIC Document ED380198)

Fenwick, L.T. (1996). A Perspective on Race Equity and Science and Math Education: Toward Making Science and Math for All. Paper presented at the Annual Conference of the Georgia Initiative in Mathematics and Science, Atlanta, GA.

Hedges, L.V., & Nowell, A. (1998). Black-White test score convergence since 1965. In C. Jencks & M. Phillips (Ed.s), The Black-White Test Score Gap (pp. 149-181). Washington, D.C.: The Brookings Institution.

Johnson, K.A. (2000). The peer effect on academic achievement among public elementary school students (Report No. CDA00-06). Washington, D.C.: The Heritage Foundation. (ERIC Document ED442916)

Miller-Whitehead, M. (1999, November). Bridging the Student Achievement Gap in Science. Paper presented at the Annual Meeting of the Mid South Educational Research Association, Point Clear, AL.

Mitchell, F.M. (1999, April). All Students Can Learn: Effects of Curriculum Alignment on the Mathematics Achievement of Third-Grade Students. Paper presented at the Annual Meeting of the American Educational Research Association, Montreal, Quebec, Canada.

Patterson, C. J., Kupersmidt, J. B., & Vaden, N. A. (1990). Income level, gender, ethnicity, and household composition as predictors of children's school-based competence. Child Development, 61, 485-494.

Peng, S. S., & Wright, D. (1994). Explanation of academic achievement of asian american students. Journal of Educational Research, 87 (6), 346-352.

Saturnelli, A. M., & Repa, J. T. (1995, April). Alternative Forms of Assessment in Elementary Science: The Interactive Effects of Reading, Race, Economic Level and the Elementary Science Specialist on Hands-On and Multiple-Choice Assessment of Science Process Skills. Paper presented at the Annual Conference of the American Educational Research Association, San Francisco, CA.

Wong, K.K., & Alkins, K.F. (1999, April). Toward Systemic Reform in High Poverty Schools: A Comparative Analysis of Two Large School Districts. Paper presented at the Annual Meeting of the American Educational Research Association, Montreal, Quebec, Canada.

Yellin, D., & Koetting, J.R. (1991). Literacy as Emancipation. Education Digest, March, 14-16.

# Table 1

# Key Group Characteristics: WASL

| | | | | | Group | | | | | | | | | | |
|---|---|---|---|---|---|---|---|---|---|---|---|---|---|---|---|
| | WASL-499 | | | | WASL-400 | | | | WASL-799 | | | | | | |
| Variable | N | Mean | SD | Min | Max | N | Mean | SD | Min | Max | N | Mean | SD | Min | Max |
| **Low Income** | | | | | | | | | | | | | | | |
| %Free Lunch | 1042 | 0.39 | 0.23 | 0.00 | 0.98 | 1057 | 0.38 | 0.23 | 0.00 | 1.00 | 426 | 0.34 | 0.20 | 0.00 | 0.95 |
| **Ethnicity** | | | | | | | | | | | | | | | |
| American Indian | 1058 | 3.22 | 7.71 | 0.00 | 100.00 | 1073 | 3.23 | 8.73 | 0.00 | 100.00 | 445 | 4.38 | 10.93 | 0.00 | 100.00 |
| Asian | 1058 | 6.66 | 8.74 | 0.00 | 63.33 | 1073 | 6.78 | 8.47 | 0.00 | 66.67 | 445 | 4.90 | 6.63 | 0.00 | 45.86 |
| African American | 1058 | 5.50 | 9.95 | 0.00 | 90.38 | 1073 | 5.61 | 10.02 | 0.00 | 95.24 | 445 | 3.84 | 8.56 | 0.00 | 86.54 |
| Hispanic | 1058 | 8.59 | 13.97 | 0.00 | 84.51 | 1073 | 9.42 | 14.76 | 0.00 | 91.43 | 445 | 8.44 | 13.69 | 0.00 | 76.81 |
| White | 1058 | 73.45 | 21.50 | 0.00 | 100.00 | 1073 | 73.82 | 21.62 | 0.00 | 100.00 | 445 | 74.19 | 21.12 | 0.00 | 100.00 |
| **WASL Scale** | | | | | | | | | | | | | | | |
| Reading | 1059 | 403.99 | 7.46 | 378.54 | 435.46 | 1074 | 407.07 | 7.67 | 381.55 | 443.60 | 444 | 392.18 | 7.78 | 364.43 | 418.35 |
| Math | 1059 | 385.77 | 15.00 | 327.95 | 436.62 | 1074 | 390.92 | 15.07 | 345.54 | 453.30 | 444 | 361.52 | 22.96 | 270.78 | 471.83 |
| Listening | 1059 | 412.61 | 17.80 | 344.15 | 471.23 | 1074 | 411.09 | 18.77 | 343.03 | 474.93 | 444 | 441.17 | 18.65 | 365.36 | 536.67 |
| Writing | 1059 | 365.35 | 19.24 | 287.53 | 435.67 | Standardized Scores Unavailable | | | | | 444 | 364.61 | 23.46 | 269.00 | 442.17 |

Note. WASL = Washington Assessment of Student Learning test. N = number of schools. SD = standard deviation. Min = minimum. Max = maximum.

# Table 2

# Key Group Characteristics: ITBS

| Variable | | | Group | | | | | | | | | | | | |
|---|---|---|---|---|---|---|---|---|---|---|---|---|---|---|---|
| | ITBS-399 | | | | | ITBS-300 | | | | | ITBS-600 | | | | |
| | N | Mean | SD | Min | Max | N | Mean | SD | Min | Max | N | Mean | SD | Min | Max |
| **Low Income** | | | | | | | | | | | | | | | |
| %FreeLunch | 1040 | 0.40 | 0.23 | 0.00 | 0.98 | 1055 | 0.38 | 0.23 | 0.00 | 0.99 | 707 | 0.35 | 0.21 | 0.00 | 0.95 |
| **Ethnicity** | | | | | | | | | | | | | | | |
| American Indian | 1062 | 3.01 | 8.43 | 0.00 | 100.00 | 1077 | 3.23 | 9.10 | 0.00 | 100.00 | 735 | 3.60 | 9.62 | 0.00 | 100.00 |
| Asian | 1062 | 6.52 | 8.30 | 0.00 | 62.34 | 1077 | 6.81 | 8.77 | 0.00 | 62.75 | 735 | 6.21 | 7.38 | 0.00 | 57.97 |
| African American | 1062 | 5.40 | 9.61 | 0.00 | 90.48 | 1077 | 5.70 | 9.99 | 0.00 | 92.86 | 735 | 4.21 | 7.84 | 0.00 | 95.83 |
| Hispanic | 1062 | 8.95 | 14.57 | 0.00 | 94.03 | 1077 | 9.74 | 15.44 | 0.00 | 90.70 | 735 | 7.46 | 12.67 | 0.00 | 90.00 |
| White | 1062 | 74.11 | 21.61 | 0.00 | 100.00 | 1077 | 71.74 | 22.84 | 0.00 | 100.00 | 735 | 74.81 | 20.58 | 0.00 | 100.00 |
| Multiracial | 1062 | 2.00 | 4.63 | 0.00 | 82.14 | 1077 | 0.87 | 2.70 | 0.00 | 30.00 | 735 | 1.68 | 4.62 | 0.00 | 45.83 |
| **ITBS Scale** | | | | | | | | | | | | | | | |
| Reading | 1062 | 185.91 | 7.98 | 159.58 | 225.54 | 1077 | 186.53 | 8.01 | 158.56 | 225.14 | 734 | 228.42 | 10.97 | 172.67 | 275.36 |
| Math | 1061 | 187.20 | 7.37 | 165.11 | 221.92 | 1077 | 188.60 | 7.68 | 164.59 | 223.52 | 734 | 230.22 | 10.50 | 178.57 | 273.25 |
| Language | 714 | 183.62 | 9.92 | 140.00 | 224.96 | 783 | 184.23 | 11.09 | 138.00 | 230.29 | 733 | 232.05 | 13.57 | 183.50 | 293.80 |
| Vocabulary | 1062 | 185.14 | 8.38 | 156.65 | 222.51 | 1077 | 185.76 | 8.38 | 152.78 | 222.56 | 734 | 226.57 | 10.48 | 161.83 | 273.50 |

Note. ITBS = Iowa Test of Basic Skills. N = number of schools. SD = standard deviation. Min = minimum. Max = maximum.

Table 3

## WASL Scores Predicted by Low Income and Percentage of White Students in Three Groups

| Outcome | Group | | | | | | | | | | | | | | |
| --- | --- | --- | --- | --- | --- | --- | --- | --- | --- | --- | --- | --- | --- | --- | --- |
| | WASL-499 | | | | | WASL-400 | | | | | WASL-799 | | | | |
| Predictor | B | Beta | R | R2-Ch | R2-Tot | B | Beta | R | R2-Ch | R2-Tot | B | Beta | R | R2-Ch | R2-Tot |
| **Reading** | | | | | | | | | | | | | | | |
| Overall Model | | | | | .55 | | | | | .53 | | | | | .53 |
| Intercept | 406.27 | | | | | 408.68 | | | | | 391.57 | | | | |
| %on Free Lunch | -19.14 | -0.60 | -0.72 | 0.23 | | -19.14 | -0.58 | -0.71 | 0.21 | | -20.01 | -0.53 | -0.69 | 0.18 | |
| %White | 0.07 | 0.21 | 0.57 | 0.03 | | 0.08 | 0.22 | 0.56 | 0.03 | | 0.10 | 0.27 | 0.59 | 0.05 | |
| **Math** | | | | | | | | | | | | | | | |
| Overall Model | | | | | .49 | | | | | .47 | | | | | .49 |
| Intercept | 389.79 | | | | | 394.63 | | | | | 359.31 | | | | |
| %on Free Lunch | -36.18 | -0.56 | -0.68 | 0.20 | | -35.70 | -0.55 | -0.67 | 0.19 | | -56.46 | -0.50 | -0.66 | 0.16 | |
| %White | 0.14 | 0.20 | 0.53 | 0.03 | | 0.13 | 0.20 | 0.52 | 0.02 | | 0.29 | 0.26 | 0.57 | 0.04 | |
| **Listening** | | | | | | | | | | | | | | | |
| Overall Model | | | | | .49 | | | | | .47 | | | | | .45 |
| Intercept | 412.10 | | | | | 411.48 | | | | | 434.01 | | | | |
| %on Free Lunch | -39.27 | -0.51 | -0.67 | 0.17 | | -41.60 | -0.51 | -0.66 | 0.17 | | -38.70 | -0.44 | -0.62 | 0.12 | |
| %White | 0.21 | 0.26 | 0.57 | 0.04 | | 0.21 | 0.24 | 0.55 | 0.04 | | 0.26 | 0.31 | 0.57 | 0.06 | |
| **Writing** | | | | | | | | | | | | | | | |
| Overall Model | | | | | .36 | Standardized Scores Unavailable | | | | | | | | | .36 |
| Intercept | 384.56 | | | | | | | | | | 371.40 | | | | |
| %on Free Lunch | -50.06 | -0.60 | -0.60 | 0.23 | | | | | | | -55.71 | -0.49 | -0.59 | 0.16 | |
| %White | 0.00 | **0.01** | 0.37 | 0.00 | | | | | | | 0.16 | 0.15 | 0.45 | 0.01 | |

Note. B = unstandardized regression coefficients. R = simple correlation between outcome and denoted variable. R²-Ch = percentage of variance in outcome variable explained by denoted variable, above and beyond remaining variable. R²-Tot = total variance explained by low income and percentage of white students as a set. All predictors significant at $p < .01$, with exception of bolded betas.

Table 4

# ITBS Scores Predicted by Low Income and Percentage of White Students in Three Groups

| Outcome | | Group | | | | | | | | | | | | | | |
|---|---|---|---|---|---|---|---|---|---|---|---|---|---|---|---|---|
| | | ITBS-399 | | | | | ITBS-300 | | | | | ITBS-600 | | | | |
| Predictor | | B | Beta | R | R2-Ch | R2-Tot | B | Beta | R | R2-Ch | R2-Tot | B | Beta | R | R2-Ch | R2-Tot |
| **Reading** | | | | | | | | | | | | | | | | |
| Overall Model | | | | | | .57 | | | | | .59 | | | | | .58 |
| Intercept | | 189.70 | | | | | 189.87 | | | | | 228.35 | | | | |
| %on Free Lunch | | -21.58 | -0.64 | -0.74 | 0.26 | | -21.59 | -0.64 | -0.75 | 0.29 | | -29.13 | -0.59 | -0.72 | 0.27 | |
| %White | | 0.06 | 0.17 | 0.56 | 0.02 | | 0.07 | 0.19 | 0.55 | 0.03 | | 0.13 | 0.27 | 0.55 | 0.05 | |
| **Math** | | | | | | | | | | | | | | | | |
| Overall Model | | | | | | .42 | | | | | .46 | | | | | .41 |
| Intercept | | 194.04 | | | | | 192.98 | | | | | 232.97 | | | | |
| %on Free Lunch | | -19.78 | -0.63 | -0.65 | 0.25 | | -19.55 | -0.60 | -0.67 | 0.25 | | -25.89 | -0.54 | -0.62 | 0.22 | |
| %White | | 0.01 | 0.04 | 0.42 | 0.00 | | 0.04 | 0.13 | 0.45 | 0.01 | | 0.08 | 0.17 | 0.43 | 0.02 | |
| **Language** | | | | | | | | | | | | | | | | |
| Overall Model | | | | | | .27 | | | | | .23 | | | | | .46 |
| Intercept | | 195.24 | | | | | 193.55 | | | | | 238.08 | | | | |
| %on Free Lunch | | -22.95 | -0.56 | -0.51 | 0.19 | | -23.00 | -0.49 | -0.48 | 0.15 | | -37.23 | -0.60 | -0.67 | 0.27 | |
| %White | | -0.03 | -0.08 | 0.28 | 0.00 | | -0.01 | **-0.02** | 0.28 | 0.00 | | 0.09 | 0.14 | 0.43 | 0.02 | |
| **Vocabulary** | | | | | | | | | | | | | | | | |
| Overall Model | | | | | | .58 | | | | | .60 | | | | | .58 |
| Intercept | | 187.17 | | | | | 188.24 | | | | | 225.87 | | | | |
| %on Free Lunch | | -21.48 | -0.60 | -0.74 | 0.23 | | -22.07 | -0.63 | -0.75 | 0.27 | | -27.23 | -0.59 | -0.72 | 0.26 | |
| %White | | 0.09 | 0.22 | 0.59 | 0.03 | | 0.08 | 0.23 | 0.57 | 0.04 | | 0.13 | 0.28 | 0.57 | 0.06 | |

Note. B = unstandardized regression coefficients. $R$ = simple correlation between outcome and denoted variable. $R^2$-Ch = percentage of variance in outcome variable explained by denoted variable, above and beyond remaining variable. $R^2$-Tot = total variance explained by low income and percentage of white students as a set. All predictors significant at $p < .01$, with exception of bolded betas.

Washington School Research Center

**Technical Report #5 – February 2003**

# Constructivist Teaching and Student Achievement: The Results of a School-level Classroom Observation Study in Washington

## Martin L. Abbott, Ph.D.

## Jeffrey T. Fouts, Ed.D.

**The Washington School Research Center (WSRC)** is an independent research and data analysis center within Seattle Pacific University. The Center began in July 2000, funded through a gift from the Bill and Melinda Gates Foundation. Our mission is to conduct sound and objective research on student learning in the public schools, and to make the research findings available for educators, policy makers, and the general public for use in the improvement of schools. We believe that sound data and appropriate data analysis are vital components for the identification of school and classroom practices related to increased student academic achievement.

Washington School Research Center
3500 188th St. S.W., Suite 328
Lynnwood, WA 98037
Phone: 425.744.0992
Fax: 425.744.0821
Web: www.spu.edu/wsrc

<table>
<tr><td>**Jeffrey T. Fouts**</td><td>**Martin L. Abbott, Ph.D.**</td><td>**Duane B. Baker, Ed.D.**</td></tr>
<tr><td>Executive Director Professor of Education</td><td>Senior Researcher Professor of Sociology</td><td>Director - School Information Services</td></tr>
</table>

# Constructivist Teaching and Student Achievement: The Results of a School-level Classroom Observation Study in Washington

A Technical Report For
The Washington School Research Center

Washington School Research Center

# Foreword

In 2000 the Bill & Melinda Gates Foundation began an education initiative in Washington State that centered on school reinvention with the goal of improving student learning. As part of that initiative third party evaluation teams have been monitoring the process and progress of reinvention and collecting various forms of data from the schools. During the 2001–02 school year one of these teams conducted an extensive classroom observation study in 34 schools to determine the degree to which "powerful teaching and learning" (also called constructivist teaching or authentic instruction) was present in the schools. The findings from that study were presented in a descriptive report and showed that this form of teaching was present in about 17 percent of the classrooms they observed. The data from that classroom observation study have been provided to the Washington School Research Center for further analysis, resulting in this technical report on the relationship of constructivist teaching to student achievement.

The findings presented here are at the same time instructive and disturbing. The relationship between student family income and student achievement is expected and a consistent finding in virtually all studies using aggregated school-level data. The strong relationship between constructivist teaching practices and student achievement is somewhat surprising, given the aggregated nature of the data used in the analyses. From a theoretical perspective the state essential learnings, WASL assessments and the theoretical model of constructivist teaching used in the observation study appear to be very complementary, and these data support that model. This finding suggests that Washington schools should consider the potential advantages of these instructional practices.

Critics of American education have claimed that children living in poverty often receive an inferior educational experience. Unfortunately, at least in this sample of schools, the relatively strong negative correlation between school-level student family income and constructivist teaching shows that students in schools with lower levels of student family income receive less intellectually demanding instruction and less instruction of the type that is a predictor of academic success than do students in schools with higher levels of family income. This finding should be concern to all of us as we work to improve education in this state.

Jeffrey T. Fouts
Executive Director

# Table of Contents

Introduction ........................................................................................................ 1

The Classroom Observation Study .................................................................. 1

The Model Tested ............................................................................................. 2

The Nature of the Data .................................................................................... 3

Findings ............................................................................................................. 4

Discussion .......................................................................................................... 6

References .......................................................................................................... 9

# Constructivist Teaching and Student Achievement: The Results of a School-level Classroom Observation Study in Washington

## Introduction

A classroom observation study was conducted in the 2001–2002 school year among a selection of Washington schools to identify the extent of constructivist teaching activity. While the findings of this study were informative in pointing out the nature and extent of the kind of teaching that occurs in schools, it was important to see whether the findings could predict school-level student achievement in schools that varied by low-income.[1] Prior research suggests that constructivist teaching (and other school-level attributes) has an impact on student achievement.[2] But can constructivist teaching predict school-level achievement beyond the effects of low-income?

## The Classroom Observation Study

The classroom observation study was part of the ongoing program evaluation of the Bill & Melinda Gates Foundation's Model Schools Initiative and Model Districts Initiative in the state of Washington. A complete description of the study and results is provided by Fouts, Brown and Thieman (2002). The study used the Teaching Attributes Observation Protocol (TAOP), which is based on a conceptual framework of constructivist teaching and learning. The TAOP contains seven lesson components and a number of indicators under each component. The content validity of the instrument was checked against the literature and existing observation instruments.

Following an extensive training period, classroom observations were conducted in 669 classrooms from 34

---

[1]Low-income was measured by the percent of students at a school who are eligible for compensatory funding.
[2]Wilson, et al., 2002.

schools over a four month period of time. The sample of schools consisted of 15 elementary, eight middle/junior high, nine high, and two technical schools. The study was designed to provide for a representative sample of classrooms drawn from social studies, mathematics, science, and language arts/ English classrooms. The number of classrooms observed at a school ranged from 6 to 54 classes, depending on the size of the school. Provisions were made for continual checks for inter-rater reliability and agreement, and the results suggest that there was a high degree of consistency in the rating process.

The general findings of this study were that strong constructivist teaching was observable in about 17% of the classroom lessons. The other 83% of the lessons observed may have contained some elements of constructivist teaching, but as many as one-half of the lessons observed had very little or no elements of constructivist teaching present. More constructivist teaching appeared to take place in alternative schools and in integrated subject matter classes than in traditional schools or subjects. There appeared to be no differences among the elementary, middle/junior high and high schools as to the degree to which constructivist practices were used.

While this classroom observation study provided a general description of the degree to which constructivist practices were employed in the schools, the relationship of this practice to student achievement was not explored. To determine the degree of that relationship is the purpose of this report.

## The Model Tested

Because of the potential overlap of low-income (LI) and constructivist teaching activity (CT) by school, we proceeded with an incremental partitioning of the variance among the study variables. The following model posits direct and indirect effects of LI on achievement (A) and direct effects of CT on A:

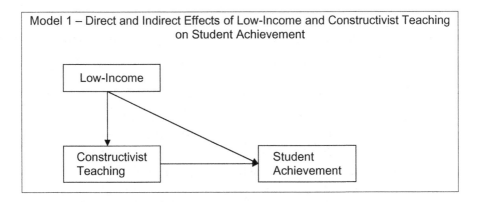

Model 1 – Direct and Indirect Effects of Low-Income and Constructivist Teaching on Student Achievement

The goal of this analysis was to understand the percent of variance in student achievement accounted for by constructivist teaching beyond that contributed by low-income. Previous studies[3] have noted the substantial impact of low-income on student achievement in Washington. This study proceeds from that research by attempting to document the incremental variance contributed by constructivist teaching.

## The Nature of the Data

Data used in this analysis were gathered from two sources. First, the classroom observation data were obtained from the researchers who conducted the observation study. The TAOP provides seven lesson component scores and a holistic score on the overall degree of constructivist instruction for each lesson. This holistic score for each classroom observed was aggregated for each school, providing a school-level constructivist teaching score. These scores provided the measure of the degree of constructivist teaching practices at the school.

The second source of data was the school-level WASL scores of 4th, 7th, and 10th grade students in Washington provided by the Office of Superintendent of Public Instruction. School scores indicated the percent of the students who passed the writing, reading, and mathematics sections of the

---

[3] For example, Abbott and Joireman, 2000.

WASL administered in 2002.[4] Schools designated "Alternative schools" were not included in the final analysis since they represent different classroom arrangements. This resulted in a final sample of 28 schools used in the following analyses.

Each of the 28 schools was assigned three achievement (writing, reading, and mathematics) scores according to the appropriate grade level of the school. In a few instances, schools that had scores for both 4th and 7th grades were assigned scores depending on the size of the grade levels. Standard scores (z) were created for the grade-level scores for greater comparability across school levels.

## Findings

Table 1 shows the descriptive statistics of the variables used in the analysis.

Table 1

Descriptive Statistics

|                              | Mean  | SD   |
|------------------------------|-------|------|
| Writing (z)                  | 0.47  | 0.83 |
| Reading (z)                  | 0.33  | 0.89 |
| Mathematics (z)              | 0.43  | 1.07 |
| Constructivist Teaching (z)  | -0.21 | 0.84 |
| Low-Income (z)               | 0.06  | 0.97 |
| N = 28                       |       |      |

Table 2 shows the zero order correlations among the study variables. An inspection of the correlations shows that low-income is correlated inversely with all of the other study variables, including achievement. This was expected given the

[4] % passing included students achieving Level 3 and 4 on reading and mathematics, and Level 4 on writing.

findings of other studies involving low-income and achievement (see earlier note). Also important to note however, is the significant inverse correlation between low-income and constructivist teaching (–.54). This is a general indication that schools that have more low-income students also have lower constructivist teaching scores. The correlations among CT and achievement variables were strong and positive.

Table 2

Correlations Among Study Variables

|   |   | 1 | 2 | 3 | 4 | 5 |
|---|---|---|---|---|---|---|
| 1 | Writing | 1 | | | | |
| 2 | Reading | 0.90 | 1 | | | |
| 3 | Mathematics | 0.86 | 0.93 | 1 | | |
| 4 | Low-Income | -0.82 | -0.89 | -0.87 | 1 | |
| 5 | Constructivist Teaching | 0.61 | 0.65 | 0.62 | -0.54 | 1 |

(All correlations are significant at or beyond the .01 level)

The second part of the analysis was the incremental partitioning of the variance according to Model 1, given the level of inter-correlations among the study variables. In this analysis, the increment of variance in achievement accounted for by constructivist teaching beyond that contributed by low-income was determined by the following formula:

$$R^2_{1.23} - R^2_{1.2} \quad (1 = A; 2 = LI; 3 = CT)$$

The first part of the formula is the entire regression model, and the second part is the regression of achievement on low-income. Applying this formula is equivalent to the squared semi-partial correlation of constructivist teaching and achievement, beyond the influence of low-income.

Table 3 shows the results of the regression analyses for average school-level writing, reading, and mathematics scores. In each case, the results for the omnibus test ($R^2$ and

F-ratio) are given, along with the contribution of CT to achievement beyond the effects of low-income. When low-income is accounted for, CT contributes between 3% and 4% of the variance in achievement. Taken together, these results indicate positive but small effects of constructivist teaching on school-level achievement beyond the contribution of low-income.

Table 3

Contribution of Constructivist Teaching to Achievement, Accounting for Low-Income

| Achievement | Overall $R^2$ (Adj.) | Overall F Ratio | Unique Contribution of CT[1] |
|---|---|---|---|
| Writing | .683 | 30.111, p < .001 | .04 |
| Reading | .824 | 64.143, p < .001 | .04* |
| Mathematics | .774 | 47.261, p < .001 | .03 |

[1]Unique contribution of CT is indicated by the squared semi-partial correlation　　* (p < .05)

## Discussion

The most notable findings in these analyses are the large correlations between the study variables. The negative correlation between school-level family income and student achievement was expected. The large positive correlations between constructivist teaching and student achievement are noteworthy, as is the negative correlation between constructivist teaching and school-level family income. The regression analyses show that constructivist teaching does predict student achievement beyond the effects of school-level family income, albeit with a greatly reduced effect.

While the unique contribution of constructivist teaching to achievement is small, there are several observations that can be made about the findings. First, school-level analyses of this sort often indicate that low-income has a dominant influence on student achievement. In an earlier study in

Washington State researchers found that "the only variables that were significant predictors of WASL scores were aggregate demographic factors" (Stecher & Chun, 2001, p. 23). This being the case, it is perhaps noteworthy that constructivist teaching predicts additional variance in achievement, given the nature of the data used in these analyses.

Second, aggregate constructivist teaching scores are more inclusive of academic subjects than aggregate WASL scores. The constructivist teaching score is a gross measure of the type of instruction occurring in a variety of classrooms, whereas WASL assessments target math, reading, and writing specifically. In this sense, the aggregated constructivist teaching scores represent an overall instructional score of several disciplines and are not limited to reading, writing, and math. One might theorize that the aggregated observation ratings in only math and language arts classrooms might be stronger predictors of the corresponding WASL scores. In this study, these analyses were not possible because of the small number of classroom observation scores in some of the math and language arts classrooms for some schools.

Third, $7^{th}$ and $10^{th}$ grade success on the WASL may be due to the cumulative effects of instruction received in prior years, rather than the type of instruction received in the present or year before the WASL assessment. Therefore, any effects of current instructional practices would be expected to have only a limited impact.

Fourth, the correlations may be affected because of a restriction of range in both the test scores and the classroom observation data. The classroom observation researchers noted that the variance within a school was much greater than among schools. This statistical limitation has the effect of reducing correlations. For these reasons, the scores might underestimate the importance of constructivist teaching, particularly given the strong zero-order correlation and the nature of the intellectual activities and student performance required for success on the WASL.

Finally, the negative correlation between constructivist teaching and student family income points out that, for whatever reason, students in schools with lower levels of student family income receive qualitatively different instruction than do students in schools with higher levels of family income. Specifically, they receive less constructivist teaching, and as measured by the TAOP, this means less intellectually demanding instruction. This finding warrants further exploration and attention.

# References

Fouts, J.T., Brown, C.J., & Thieman, G. (2002). *Classroom instruction in Gates grantee schools: A baseline report.* Seattle, WA: The Bill & Melinda Gates Foundation.

Joireman, J., & Abbott, M. (2001). *The relationships between the Iowa Test of Basic Skills and the Washington Assessment of Student Learning in the State of Washingon* [ERIC Document TM033308]. Lynnwood, WA: Washington School Research Center, Seattle Pacific University.

Stecher, B., & Chun, T. (2001). *The effects of the Washington education reform on school and classroom practice, 1999–2000.* Santa Moncia, CA: RAND.

Wilson, B., Abbott, M., Joireman, J., & Stroh, H. (2002). *The relations among school environment variables and student achievement: A structural equation modeling approach to effective schools research. Technical Report #4.* Lynnwood, WA: Washington School Research Center, Seattle Pacific University.

# REFERENCES

Abbott ML, Baker D, Smith K, Trzyna T. *Winning the Math Wars: No Teacher Left Behind*. Seattle: University of Washington Press; 2010.

Abbott ML, Joireman J, Stroh H. (Washington School Research Center, Seattle Pacific University, Seattle, WA). The influence of district size, school size and socioeconomic status on student achievement in washington: a replication study using hierarchical linear modeling. Technical Report 3, 2002 November. *http://www.spu.edu/orgs/research/ WSRC%20HLM%20District%20Size%20Final%2010-2-02.pdf*.

Aiken LS, West SG. *Multiple Regression: Testing and Interpreting Interactions*. Newbury Park, CA: Sage; 1991.

Ayres I. *Super Crunchers*. New York: Bantam; 2007.

Baron RM, Kenney DA. The moderator–mediator variable distinction in social psychological research: Conceptual, strategic, and statistical considerations. *Journal of Personality and Social Psychology* 1986; 51: 1173–82.

Berger PL. *Invitation to Sociology*. New York: Doubleday; 1963.

Bickel R. *Multilevel Analysis for Applied Research*. New York: Guilford Press; 2007.

Bickel R, Howley C. The influence of scale on student performance: A multi-level extension of the Matthew Principle. *Education Policy Analysis Archives* 2000; 8 (22).

Brown TA. *Confirmatory Factor Analysis for Applied Research*. New York: Guilford Press; 2006.

Byrne BG. *Structural Equation Modeling with AMOS: Basic Concepts, Applications, and Programming*, 2nd ed. New York: Routledge; 2010.

Cohen J, Cohen P, West SG, Aiken LS. *Applied Multiple Regression/Correlation Analysis for the Behavioral Sciences*, 3rd ed. Mahwah, NJ: Erlbaum; 2003.

Dostoevsky F. *Crime and Punishment*. New York: Knopf; [1866] 1993.

Friedman T. *The World Is Flat: A Brief History of the Twenty-first Century.* New York: Farrar, Straus and Giroux; 2005.

*Guiding Principles for Evaluators. American Evaluation Association website (http://www.eval.org/GPTraining/GPTrainingOverview.asp)*

Green SB, Salkind NJ. *Using SPSS for Windows and Macintosh*, 5th ed. Boston: Pearson; 2008.

Grimm LG, Yarnold PR, eds. *Reading and Understanding Multivariate Statistics.* Washington, DC: American Psychological Association; 1998.

Hosmer DW, Lemeshow S. *Applied Logistic Regression.* Hoboken, NJ: Wiley; 2000.

Huff D. *How to Lie with Statistics.* New York: Norton; 1993.

Judd CM, McClelland GH. *Data Analysis: A Model Comparison Approach.* New York: Harcourt Brace Jovanovich; 1989.

Levitt SD, Dubner SJ. *Freakonomics.* New York: Harper Collins; 2005.

Lofland J, Lofland LH. *Analyzing Social Settings*, 3rd ed. Boston: Wadsworth; 1995.

McLachlan GJ. *Discriminant Analysis and Statistical Pattern Recognition.* Hoboken, NJ: Wiley; 1992.

Ouchi W. *Theory Z: How American Business Can Meet the Japanese Challenge.* Menlo Park, CA: Addison Wesley; 1981.

Pedhazur EJ. *Multiple Regression in Behavioral Research*, 3rd ed. Fort Worth: Harcourt; 1997.

Raudenbush SW, Bryk AS. *Hierarchical Linear Models*, 2nd ed. Thousand Oaks, CA: Sage; 2002.

Raudenbush SW, Bryk AS, Cheong YF, Congdon R. *HLM 6: Linear and Nonlinear Modeling.* Lincolnwood, IL: SSI, Inc.; 2004.

*SPSS® for Windows, Rel. 17.0.2.* Chicago: SPSS®, Inc.; 2009.

Tabachnick BG, Fidell LS. *Using Multivariate Statistics*, 5th ed. Boston: Pearson; 2007.

# INDEX

Adjusted $R^2$, 103, 118
Aggregated data, 54–56, 83
ANOVA (analysis of variance)
   in regression results, 103, 118

*b* coefficient, 92, 155
Beta
   standardized (β), 118, 158
   unstandardized, 118, 158
Bivariate correlation, *see*
   Correlation, bivariate
   correlation
Bivariate regression, *see* Regression

Categorical data, 55–56, 83
Causality, 13–14
   versus correlation, 50–51
Causal direction, 69–70
Centered leverage values, *see* SPSS,
   centered leverage values
Classroom observation data,
   28–29
Cleaning data, 121–136
Coding, 185–204
   contrast, 186, 204
   dummy, 185–203, 204
   effect, 186, 205
   orthogonal, 186

subvariables created in dummy
   coding, 193–202, 205
Coefficient of determination,
   *see* Correlation, effect size
Confidence intervals, *see* Regression,
   confidence intervals
Control variables, 139–140, 149.
   *See also* Mediator variables
Cook's distance value, *see* SPSS,
   Cook's distance values
Correlation, 47–86. *See also*
   Pearson's *r*
   assumptions of, 79–81
   bivariate correlation, 83
   curvilinear, 48–49, 81, 83
   direction, 47–49, 70–71
   effect size (coefficient of
   determination), 61, 67–69, 83.
   *See also* $R^2$
   influences on, 71–73
   Kendall's tau-b, 64, 74, 76–78
   matrix, 58–60, 68
   nature of, 47–49
   non-linear correlation, 78, 81
   prediction, 49–50
   scatter diagram, 47–49
   Spearman's rho, 64, 74–78
   vs. causation, 50–51

*The Program Evaluation Prism: Using Statistical Methods to Discover Patterns*,
by Martin Lee Abbott
Copyright © 2010 John Wiley & Sons, Inc.

Curve estimation, *see* SPSS, curve estimation
Curvilinear relationship, 48–49, 81, 83

DfBeta, *see* SPSS, DfBeta
Dichotomous variables, 55–56. *See also* Categorical data
Dichotomy, 55
Discovery learning, 4, 24, 225–226
Discriminate analysis, 232
Distance statistics, *see* SPSS, distance statistics

Ecological fallacy, 54–55, 83
Effect size, 20–21, 29, 60–63. *See also* Statistical significance
Evaluation prism, 10, 81–82
Evaluation research, 29
program evaluation, 1–5, 9–11
Evaluation standards, 30
Explanation versus prediction, 87–88, 153–154
Extreme scores, *see* Outlier scores

F tests, 93
Factor analysis, 63–64, 83, 232. *See also* Principal Components Analysis
False dichotomies, 203–204, 205
Fidelity, 13, 21, 30
Fixed effects modeling, 111. *See also* Random effects modeling

Heteroscedasticity, 119
Hierarchical analyses, 166–167, 181
Hierarchical linear modeling, 54, 229–230
Homoscedasticity, 110, 119

Influence statistics, *see* SPSS, influence statistics
Interaction, 207–222
categorical variables, 216–221
centering, 209–210, 222

continuous variables, 209–216
disordinal, 208, 222
ordinal, 208, 222
product terms, 209, 222
simple slopes, 222
Interval data, 73–74
assumptions of correlation, 79–81

Kendall's tau-b, *see* Correlation, Kendall's tau-b

Line of best fit, 88–91, 119
Linear regression, *see* Regression
Logistic regression, 3, 184, 231–232

Median split, 187–188, 205
Mediator variables, 140–142, 149–150. *See also* Control variables
Multicollinearity, 171–174, 181
tolerance, 173–174, 182
VIF, 173–174, 182
Multilevel regression, *see* Hierarchical linear modeling
Multinomial logistic regression (MLR), 232
Multiple coefficient of determination, 138, 154
Multiple correlation, 137–149, 152
Multiple regression, 150, 153–181
assumptions of, 171–174
Multivariate extreme scores, *see* Outlier scores, multivariate extreme scores

Nonlinear regression, 111–116
Non-parametric correlation procedures, 83

Omnibus test, 119, 155
Order of entry, 181
Outlier scores, 73, 84, 121–136
multivariate extreme scores, 124–128
univariate extreme scores, 122–124

Part correlation, *see* Semipartial (part) correlation
Partial correlation, 142–147, 150
Path analysis, 150, 230–231
Pearson's correlation, 84. S*ee also* Pearson's *r*
Pearson's *r*
   computing, 78–79
   interpreting, 56–57
   interval-level data, 80
   strength and direction, 51–54
Predicted values of *Y*, *see* Regression, predicted values of *Y*
Prediction versus explanation, *see* Explanation versus prediction
Prediction, 49–50
Principal components analysis, 63–64. *See also* Factor analysis

Quadratic models, 113–116. *See also* SPSS, curve estimation
Qualitative methods, 17–18, 30
Quantification, 14–17, 30
Quasi-experimental design, 12–14, 139–142, 150

$R^2$, 96. *See also* Adjusted $R^2$
   change statistic, 119
   coefficient of determination, 61
   explained variance, 96
   multiple coefficient of determination, 138
Random effects modeling, 111. S*ee also* Fixed effects modeling
Regression, 87–120
   assumptions of, 110–111
   confidence intervals, 94, 118
   formula, 91–93
   predicted values of *Y* (*Y'*), 89–91, 92
   regression line, *see* Line of best fit
   residual error, 92
   standard error of estimate (SEest), 93, 119
   *Y* intercept, 91, 92

Residual scores, 119
Residuals, 94–98
Restricted range, 71–73, 84

Scatter diagram, 84. *See also* SPSS, scatter diagram option
Semipartial (part) correlation, 147–150
Skewness, 123, 136
Slope, 92–93. *See also* *b* coefficient
Spearman's rank order correlation, 84
Spearman's rho, *see* Correlation, Spearman's rho
SPSS
   analysis functions, 41
   backward order of entry, 171
   centered leverage values, 129–130, 135
   contingency coefficient, 74
   Cook's distance values, 130–131, 135
   correlation, 64–67
   curve estimation, 81, 111–116, 119
   DfBeta, 131–134, 135
   distance statistics, 128–131, 135
   forward order of entry, 171
   general features, 32–34
   graphing functions, 41–46
   influence statistics, 131–134, 135
   management functions, 34–41
   merge, 37–41
   overview, 31–46
   partial correlation, 144–147
   Phi coefficient, 74
   $R^2$ change, 164, 165
   reading and importing data, 34
   regression, 98–110
   scatter diagram option, 42–46
   sort, 34–35
   split file, 35
   standardized residuals, 125–128, 136

SPSS (*cont'd*)
  statistical significance, 58, 60
  studentized residuals, 128, 136
  *t* test, 120, 155, 165
  transform/compute, 36
  unstandardized residuals, 125–128,
    136
Squared part correlation, 148–149,
  164–165
Standard error of estimate (SEest),
  *see* Regression, standard error
  of estimate
STAR Classroom Observation
  Protocol™, 29, 84–86
Statistical significance, 20–21, 30,
  57–60
  practical significance, 20. *See also*
    Effect size
Stepwise regression, 167–171,
  181

Structural equation modeling, 150,
  230–231
Sum of squares (SS), 95–98, 119

*t* test, *see* SPSS, *t* test
TAGLIT, 27–28
Tolerance, *see* Multicollinearity,
  tolerance

Univariate extreme scores, *see*
  Outlier scores, univariate
  extreme scores

Variance inflation factor (VIF).
  *See also* Multicollinearity, VIF

Y intercept, *see* Regression, Y
  intercept

Zero-order correlation, 182

CPSIA information can be obtained
at www.ICGtesting.com
Printed in the USA
LVOW04s1149261215

467723LV00009B/60/P

9 780470 579046